FENCES, GATES

AND

BRIDGES

A PRACTICAL MANUAL

GEORGE A. MARTIN

THREE HUNDRED ILLUSTRATIONS

Alan C. Hood & Company, Inc.

CHAMBERSBURG, PENNSYLVANIA

Fences, Gates and Bridges

This book has been produced in the United States of America. It is published by Alan C. Hood & Co., Inc., Chambersburg, Pennsylvania 17201

Library of Congress Cataloging-in-Publication Data

Freeman, Castle.

Fences, gates, and bridges : a practical manual / edited by George A. Martin.
    p.   cm.
    Originally published: New York : O. Judd, 1887. With a new foreword.
    Includes index.
    1. Fences—Design and construction. 2. Gates—Design and construction. 3. Bridges—Design and construction. I. Martin. George A., d. 1904.
TH4965.F46 1992
631.2'7—dc20   92-18206
                                                                  CIP

ISBN 0-911469-08-7

10 9 8 7

Foreword
By Castle Freeman, Jr.
Contributing Editor
The Old Farmers Almanac

When FENCES, GATES, AND BRIDGES was a new book, Grover Cleveland was President of the United States, of which there were forty. The population of the country numbered a little less than 63 million, or about a quarter of its size in 1990. Better than forty percent of the American people lived on farms and worked in or were supported by farming in one way or another. Most of them would have found this volume familiar and useful. For its publisher, the Orange Judd Company of New York, then one of the largest agricultural publishers in the land, seems to have pitched FENCES, GATES, AND BRIDGES dead at the middle of its market.

To me at least, there is a particular farm whose image takes shape in these pages. It is by no means a hardscrabble place of the kind Huck Finn visited in Arkansas, but neither is it a gentleman's manor of the Hudson Valley. Rather it is a successful middle-sized freehold, prosperous but simple, where the owner, although he can and does hire helpers, is not too grand to work alongside them, and where the farm's economy is ample to support intelligent operation but admits no frills. Over and over again, the virtues claimed for the fences and other structures here are the same: they are strong, they will last, they are cheap. The farmer in this book doesn't expect to get his work done without putting in plenty of sweat, but he counts his money carefully and he knows well the value of his

(3)

time. Every page in this volume suggests a time and place where labor and materials are abundant, cash scarce.

The author, George A. Martin, of whom I can learn nothing except that he may have been pseudonymous, is like his book and like the sound, well-run farm his book evokes. He is plain spoken and brief, he comes right to the point, and he has an admirable way of stating unmistakably how parts go together. With his skill as a clear writer, the author also knows his subject—not only fence building but the farm and its requirements. He knows the stock: the unconfinable pig, the dexterous cow. He knows his materials, especially wood. I count twenty-one species of tree in the text, each especially suited for a particular application, He knows the value of work well done, done to last, and he aims to give value himself, in authorship as in the building of simple, necessary structures.

It is difficult—it is perhaps impossible—in reading FENCES, GATES, AND BRIDGES to avoid comparing the virtues and values of the farm it implies with our farms, and indeed with the whole life of our nation today. Whatever the result of that comparison might be, however, I think there is no doubt that the author, and his hypothetical farmer, would have thought it idle. The practical question is, what can you do with this book? And the practical answer is, everything its author intended. After all, a fence is still a fence. The ways of constructing, supporting, bracing, and repairing it are today what they were then. Mr. Martin's rules still apply, his tricks still work.

Fences, gates, and bridges—simple things. If you have need of one on whatever in your life corresponds to Mr. Martin's thrifty farm, and if your own economy is in anything like his spirit, then you can do a lot worse than study his book.

CASTLE FREEMAN, JR.
Contributing Editor
The Old Farmers Almanac
Newfane, Vermont
November 1991

TABLE OF CONTENTS.

(5)

FENCES, GATES AND BRIDGES.

CHAPTER I.

RAIL AND OTHER PRIMITIVE WOOD FENCES.

VIRGINIA RAIL FENCE.

The zigzag rail fence was almost universally adopted by the settlers in the heavily timbered portions of the country, and countless thousands of miles of it still exist, though the increasing scarcity of timber has brought other styles of fencing largely into use. Properly built, of good material, on a clear, solid bed, kept free from bushes and other growth to shade it and cause it to rot, the rail fence is as cheap as any, and as effective and durable as can reasonably be desired. Good chestnut, oak, cedar, or juniper rails, or original growth heart pine, will last from fifty to a hundred years, so that material of this sort, once in hand, will serve one or two generations. This fence, ten rails high, and propped with two rails at each corner, requires twelve rails to the panel. If the fence bed is five feet wide, and the rails are eleven feet long, and are lapped about a foot at the locks, one panel will extend about eight feet in direct line. This takes seven thousand nine hundred and twenty rails, or about eight thousand rails to the mile. For a temporary fence, one that can be put up and taken down in a

(7)

short time, for making stock pens and division fences, not intended to remain long in place, nothing is cheaper or better. The bed for a fence of this kind should not be less than five feet across, to enable it to stand before the wind. The rails are best cut eleven feet long, as this makes a lock neither too long nor too short ; and the forward end of each rail should come under the next one that is laid. The corners, or locks, as they are called, should also be well propped with strong, whole rails, not with pieces of rails, as is often done. The props should be set firmly on the ground about two feet from the panel, and crossed at the lock so as to hold each other, and the top course of the fence firmly in place. They thus act as braces to the fence, supporting it

Fig. 1.—VIRGINIA ZIGZAG FENCE COMPLETE.

against the wind. Both sides of the fence should be propped. The top course of rails should be the strongest and heaviest of any, for the double purpose of weighting the fence down, and to prevent breaking of rails by persons getting upon it. The four courses of rails nearest the ground should be of the smallest pieces, to prevent making the cracks, or spaces between the rails, too large. They should also be straight, and of nearly even sizes at both ends. This last precaution is only necessary where small pigs have to be fenced out or in, as the case may be. The fence, after it is finished, will have the appearance of figure 1, will be six rails high, two props at each lock, and the worm will be crooked enough to stand any wind, that will not prostrate crops, fruit trees, etc. A straighter worm than this will be easy to blow down or push over. The stability of this sort of fence depends very largely on the manner of placing the props, both as to the distance of the foot of the prop rail from the fence panel, and the way it is locked at the corner.

LAYING A RAIL FENCE.

It is much better, both for good looks and economy, to have the corners of a rail fence on each side in line with each other. This may be accomplished by means of a very simple implement, shown in figure 2. It consists of a small pole, eight feet long, sharpened at the lower end. A horizontal arm of a length equal to half the width of the fence from extreme outside of corners, is fastened to the long pole at right angles, near the lower end. Sometimes a sapling may be found with a limb growing nearly at right angles, which will serve the purpose. Before beginning the fence, stakes are set at intervals along the middle

Fig. 2.   of the line it is to occupy. To begin, the gauge, as shown in figure 2, is set in line with the stakes, and the horizontal arm is swung outwardly at

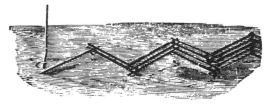

Fig. 3.—THE FENCE BEGUN.

right angles to the line of fence. A stone or block to support the first corner is laid directly under the end of the horizontal arm, and the first rail laid with one end

resting on the support. In the same way the next corner
and all others are laid, the gauge being moved from
corner to corner, set in the line of fence, and the arm
swung alternately to the right and left.

STAKING AND WIRING.

A neater and more substantial method of securing the
corners of a worm fence is by vertical stakes and wires, as
shown in the accompanying illustrations. When the
lower three rails are laid, the stakes are driven in the

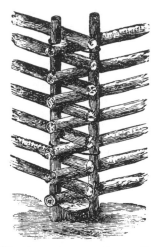 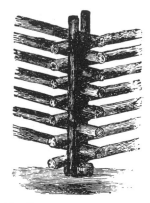

Fig. 4.—STAKES IN "LOCK."   Fig. 5.—STAKES IN ANGLES.

angles close to the rails, and secured by a band of annealed wire. The work of laying the rails proceeds, and
when within one rail of the top, a second wire band is
put in place. Or the upper wire may be put on above
the top rail. Annealed wire is plentiful and cheap.

A FENCE OF "STAKES AND RIDERS."

A very common method with the "worm" or "Virginia" rail fence is to drive slanting stakes over the corner in saw-horse style, and lay the top rail into the angle

Fig. 6.—A STAKE AND RIDER FENCE.

thus formed. The stakes, resting on the rails and standing at angle, brace the fence firmly. But the feet of the stakes extending beyond the jagged corners formed by the ends of the rail are objectionable. This is remedied in part by putting the stakes over the middle of the panel —at considerable distance apart—and laying in them long poles horizontally. In this case the stakes should be set at such an angle as to prevent their moving sidewise along the top rail, which should be a strong one. These stakes and long riders are frequently used to raise the hight of low stone walls. Figure 6 shows a fence nearly all composed of stakes and riders, which is straight and requires fewer rails than a worm fence. First, crotched stakes, formed by the forks of a branching tree limb, a foot or more long, are driven a foot or so into the ground at a distance apart corresponding to the length of poles used. The bottom poles are laid into these, and two stakes, split or round poles, are driven over these and the next poles laid in. Then two more stakes and another pole, and so on as high as the fence is required. This will answer for larger animals, and be strong and

Fig. 7.—A POLE FENCE.

not expensive. For swine, and other small live-stock, the crotch stakes may be replaced by blocks or stones, and the lower poles be small and begin close to the ground.

A POLE FENCE.

A fence which is cheaply constructed in a timbered region, and calls for no outlay whatever, besides labor, is

Fig. 8.—WITHE.   Fig. 9.—WITHE IN PLACE.

illustrated at figure 7. The posts are set in a straight line, having previously been bored with an inch augur to

receive the pins. When they are set, the pins are driven
diagonally into the posts, and the poles laid in place. It
would add much to its strength, if the poles were laid so
as to "break joints." A modification of this fence is
sometimes made by using withes instead of pins to hold
the poles in place. The withe is made of a young sapling or slender limb of beech, iron-wood, or similar tough
fibrous wood, with the twigs left on. This is twisted
upon itself, a strong loop made at the top, through which
the butt is slipped. When in place, the butt end is
tucked under the body of the withe.

FENCES FOR SOIL LIABLE TO HEAVE.

The main point in such a fence is either to set the posts

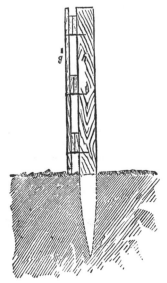

Fig. 10.—END VIEW OF FENCE.

and place a pin through them near the bottom, so that
the frost may not throw them out, or to so attach the

boards that the posts may be re-driven, without splitting them, or removing the rails from the fence. The latter is, perhaps, the best plan, and may be accomplished in

Fig. 11.—SIDE VIEW OF FENCE.

several ways, the most desirable of which is shown in figures 10 and 11. The post, h, is driven in the usual manner, when a strip of board, g, is fastened to it by three or four spikes, depending upon the hight of the

Fig. 12.—FENCE WITH IRON HOOKS.

fence. A space just sufficient to insert the ends of boards a, e, figure 11, is left between the post and outside strip, the ends of the boards resting upon the spikes. Many

miles of this fence are in use. It looks neat ; besides
any portion is easily removed, making a passage to and
from the field. A new post is easily put in when required,
and any may be re-driven when heaved by the frost.

Where iron is cheap, a rod about three-eighths of an
inch in diameter is cut in lengths of about seven and a
half inches ; one end is sharpened, while the opposite
end, for three inches, is bent at right angles. After the
boards are placed in position, the hooks should be driven
in so that they will firmly grasp the boards and hold
them in place. The general appearance of the finished
fence is shown in figure 12, and is one adapted to almost any locality.

A much better method is to fasten the boards temporarily in place, and then bore a half inch hole through

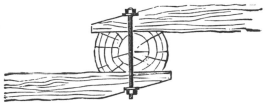

Fig. 13.—HORIZONTAL SECTION.

both boards and the post, into which a common screw
bolt is then inserted and the nut screwed on firmly. The
two ends should, however, be put on opposite sides of the
post. One bolt thus holds the ends of both boards firmly to the post, as shown in figure 13. With this style of
fence, old rails or round poles may be used instead of
boards.

OTHER PRIMITIVE FENCES.

In the heavily timbered parts of the country, where
the settlers a few years ago were making farms by felling
and burning the huge pine trees, a fence was constructed

like the one shown in figure 14.   Sections of trees, about
four and a half feet long and often as thick, were placed
in line and morticed to receive from three to five rails.

Fig. 14.—LOG POSTS.

This style of fence could be used by the landscape gar-
dener with fine effect for enclosing a park or shrubbery.

In the same regions, when a farmer has pulled all the
stumps from a pasture that slopes toward the highway,

Fig. 15.—STUMP FENCE.

the stumps may be placed in line along the road with
the top ends inside of the field.   The gaps between
where the stumps can not be rolled close together, are

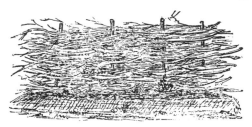

Fig. 16.—WICKER FENCE.

filled with brushwood.   A portion of this fence is shown
in figure 15.

Where other material is costly, or not to be obtained,

the wicker fence, constructed of stakes and willows, is much used. In the far West it is to be seen in every town, generally built on a small embankment of earth from one to two feet deep. In this climate, with occasional repairs, it lasts from ten to fifteen years. Figure 16 shows the style of construction.

Throughout the forest regions is found the staked and ridered brush growing on the line where the fence is

Fig. 17.—BRUSH FENCE.

constructed. Figure 17 illustrates a few rods of brush fence—such fencing being met with in our Southern States.

CHAPTER II.

STONE AND SOD FENCES.

HOW A STONE WALL SHOULD BE BUILT.

To build a stone wall, some skill is required. The foundation should be dug out a foot deep, and the earth

Fig. 18.—WELL LAID WALL.

thrown upon each side, which serves to turn water from the wall. Large stones are bedded in the trench, and long stones placed crosswise upon them. As many whole stones as possible should be used in this place. The stones are then arranged as shown in the engraving, breaking joints, and distributing the weight equally. Any small spaces should be filled with chips broken off in dressing the larger stones, so as to make them fit snugly. As it is a work that will last a century, it is worth doing well.

BUILDING A STONE FENCE.

A permanent stone fence should be built from four to five feet high, two feet wide at the base and one foot at the top, if the kind of stones available allow this construction. If a higher fence is desired, the width should be correspondingly increased. The surface of the soil along the line of the fence should be made smooth and as nearly level as possible. The hight will depend upon the situation, the animals, the smoothness of the wall (whether sheep can get foot-holds to climb over), and the character of the ground along each side. If the earth foundation be rounded up previously, sloping off to an open depression or gully, less hight will be needed. Such

Fig. 19.—LAYING UP A STONE FENCE.

an elevation will furnish a dry base not heaved by frost like a wet one. Without this, or a drain alongside or under the wall, to keep the soil always dry, the base must be sunk deeply enough to be proof against heavy frosts, which will tilt and loosen the best laid wall on wet soil. The foundation stones should be the largest; smaller stones packed between them are necessary to firmness. The mistake is sometimes made of placing all the larger stones on the outside of the wall, filling the center with small ones. Long bind-stones placed at frequent intervals through the wall add greatly to its strength. The top of the fence is most secure when covered with larger close-fitting, flat stones. The engraving shows a wooden frame and cords used as a guide in building a substantial stone fence. Two men can work together with mutual advantage on opposite sides of the stone wall.

TRUCK FOR MOVING STONES.

The small truck (figure 20) is not expensive, and may be made to save a great amount of hard lifting in building a stone wall. It is a low barrow, the side bars forming

Fig. 20.—TRUCK FOR STONE.

the handles like a wheelbarrow. It rests upon four low iron wheels. A broad plank, or two narrow ones, are laid with one end against the wall and the other resting on the ground. A groove is cut at the upper end for the wheels to rest in. The stone is loaded on the truck, moved to the place, and pushed up the plank until the wheels fall into the groove, when, by lifting on the handles, the stone is unloaded.

REINFORCING A STONE WALL.

A stone wall which affords ample protection against sheep and hogs, may be quite insufficient for horses and cattle. The deficiency is cheaply supplied in the manner indicated by the illustration, figure 21.　Round poles or

Fig. 21.—STONE WALL REINFORCED.

rails are used, and if the work is properly performed, the fence is very effective.

A COMPOSITE FENCE.

The fence illustrated at figure 22 is quite common in some parts of New England.　A ridge is thrown up by

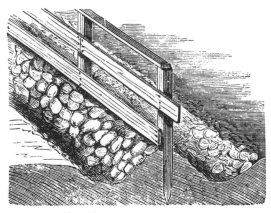

Fig. 22.—COMPOSITE FENCE.

back-furrowing with a plow, and both that and the ditches finished by hand with a shovel.　Light posts are easily driven through the soft earth, and a board fence, only three boards high, made in the usual manner. Then the stones, as they are picked up in the field, are hauled to the fence and thrown upon the ridge. This clears the field, strengthens the ridge, prevents the growth of weeds, and assists in packing the earth firmly around the bottom of the posts.

A PRAIRIE SOD FENCE.

A sod fence, beside its other value, is a double barrier against the prairie fires which are so sweeping and destructive to new settlers, if unobstructed, for a wide strip is

Fig. 23.—SOD CUTTER.

cleared of sods, the fence standing in the middle of it. A very convenient implement for cutting the sod is shown at figure 23. It is made of planks and scantling, the method of construction being clearly shown. The cutting disks are four wheel-coulters from common breaking plows, all attached to an iron shaft sixteen inches apart. They are set to cut three or four inches deep. This is run three times along the line of the fence, making nine cuts, the cutters being held down by a man riding on the rear of the apparatus. Then with a breaking plow one furrow is turned directly in the line of the fence, completely inverting the sod, the team turned to the right, and a second or back furrow is inverted on top of the first. Additional furrows are cut, diminishing in width to five or six inches on the outer side, as shown in the diagram, figure 24. After the two inner sods are turned, the rest are carried by hand, wheelbarrow or a truck, (figure 20), and laid on the sod wall, care being used to "break joints" and to taper gradually to the

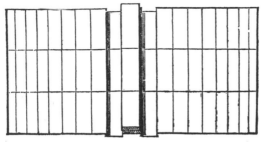

Fig. 24.—THE SOD CUT.

top. If a more substantial fence is wanted, a strip thirty-two inches wide may be left as a part for the fence, the first two furrows inverted upon the uncut portion, so that their edges just touch. The sod fence is then continued to the summit just twice as thick as it would be by the process just described. After the fence is laid, a deep furrow should be run on each side, throwing the earth against the base of the fence. A very effective and cheap fence is made by laying up a sod "dyke," as above described, three feet high, then driving light stakes along the summit, and stringing two strands of barbed wire to them.

CHAPTER III.

BOARD FENCES.

BUILDING BOARD FENCES.

In building a board fence, always start right, and it will be little trouble to continue in the same way. Much of the board fencing erected is put together very carelessly, and the result is a very insecure protection to the field or crops. A fence-post should be set two and a half or three feet in the ground, and the earth should be packed around it as firmly as possible. For packing the

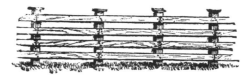

Fig. 25.—PROPERLY CONSTRUCTED BOARD FENCE.

soil there is nothing better than a piece of oak, about three inches square on the lower end, and about six feet long, rounded off on the upper part to fit the hands easily. Properly used, this instrument will pack the soil around a post as it was before the hole was dug. In putting on fence boards, most builders use two nails on the ends of each board, and one in the middle. Each board should have at least *three* nails at the ends, and *two* in the middle, and these nails should never be less than tenpennys. Smaller nails will hold the boards in place for awhile, but when they begin to warp, the nails are drawn out or loosened, and the boards drop off. This will rarely be the case where large nails are used, and a much stiffer fence is secured. Many fence builders do not cut off the tops of the posts evenly, but this should

(24)

always be done, not only for the improvement that it makes in the looks of the fence ; but also for the reason that there should always be a cap put on, and to do this, the posts must be evened. The joints should always be " broken," as is shown in the engraving, figure 25, so that in a four-board fence but two joints should come on each post. By this means more firmness and durability is secured, there being always two unbroken boards on each post to hold it in place, preventing sagging. On the face of the post immediately over where the rails have been nailed on, nail a flat piece of board the width of the post and extending from the upper part of the top rail to the ground.

Figure 26 shows a slight modification, which consists in setting the posts on alternate sides of the boards, securing additional stability. The posts are seven feet long, of

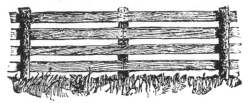

Fig. 26.—A DURABLE BOARD FENCE.

well seasoned red cedar, white oak, chestnut, or black locust, preference being accorded to order named. The boards are sixteen feet long, fastened with ten-penny steel fence nails. The posts for a space of two and a half feet from the lower end are given a good coat of boiled linseed oil and pulverized charcoal, mixed to the consistency of ordinary paint, which is allowed to dry before they are set. When the materials are all ready, stretch a line eighteen inches above the ground, where it is proposed to build the fence. Dig the post holes, eight feet apart from centers, on alternate sides of the line. The posts are set with the faces inward, each half an inch from the line, to allow space for the boards. Having set the posts, the boards of the lower course are nailed on. Then, for the first length, the second board from the bottom and the top board are only eight feet long, reaching to the first post. For all the rest the boards are of the full length, sixteen feet. By this means they " break joints." After the boards are nailed on, the top of the posts are sawed off slanting, capped, if desired,

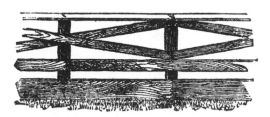

Fig. 27.—A NEAT FARM FENCE.

and the whole thing painted. A good coat of crude petroleum, applied before painting, will help preserve the fence, and save more than its cost in the paint needed.

We see another style of board fence now and then that is rather preferable to the ordinary one ; it looks better than the old straight fence. It saves one board to each length ; and by nailing on the two upper boards, as shown in the illustration, figure 27, great extra strength is given. These boards not only act as braces, but ties also, and a fence built on well set posts, and thoroughly nailed, will never sag or get out of line until the posts rot off.

FENCES FOR LAND SUBJECT TO OVERFLOW.

The fence illustrated in figures 28, 29 and 30 has posts the usual distance apart, which are hewed on the front side, and on this are nailed three blocks, three by four inches thick and six inches long ; the first one, with its

top just level with the ground, the second one, ten inches
in the clear above, and the third one, four inches less
than the desired height of the fence, measuring from

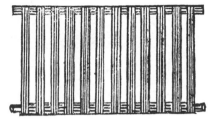

Fig. 28.—PANEL.

the top of the first block. After the panel is put in place,
the rounded ends resting on the bottom blocks, nail a
piece of board one and one-half by six inches on the
blocks, as shown in the illustrations. This board must
project four inches above the upper block, forming with
it the rest and catch for the top framing piece of the panel. The panel is made of a top and bottom piece of three

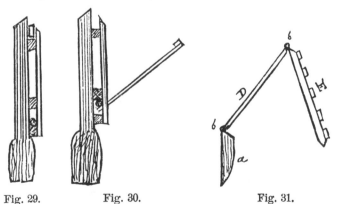

Fig. 29.   Fig. 30.   Fig. 31.

by four scantling, on which are nailed palings. The top
piece is left square, and projects three inches on each
side, but on the bottom piece the projections are cut
round, so as to turn in the slot. The water will raise the
panel up out of the upper catch, allowing it to fall down,

as seen at figure 30, so as to offer no obstruction to the
water, nor will it catch drift, as fences hung from the top

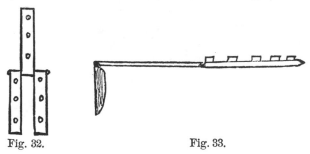

Fig. 32.   Fig. 33.

do.   Figures 31 to 35 represent a fence made somewhat
like the trestle used for drying clothes.   The posts are

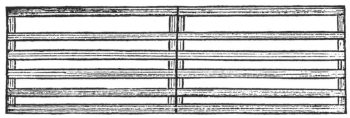

Fig. 34.

the usual distance apart, but only extend a few inches
out of the ground, just sufficient to nail a hinge upon.

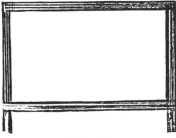

Fig. 35.

They must, however, be wide enough to admit of nailing
two hinges on each post.   The fence consists of two
parts—*E* in figure 31 represents a cross-section of the

fence proper, two panels of which are seen in figure 34; *D* represents the back part of the fence, a section of which is shown in figure 35 ; *a* in figure 31 is the post and *b b* the hinges. The panel, *E*, should always slope with the current of the stream, that the water rushing against it will place it in the position shown by figure 33, lying flat on the ground, and out of the way of both water and drift. The hinges may be ordinary strap kind, which can be bought very cheap by the dozen, or they may be made of heavy iron hoop doubled, as shown at figure 32, which can be made in any blacksmith shop.

A FENCE BOARD HOLDER.

Figure 36 shows a contrivance for holding fence boards against the posts, at the right distances apart when nail-

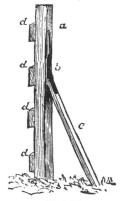

Fig. 36.   Fig. 37.—FENCE BOARD HOLDER.

ing. A two and a half by two and a half inch piece of the desired length is taken for the upright, *a*. About its center is hinged the brace, *c*. A strap hinge, *b*, or a stout piece of leather for a hinge, will answer. Blocks or stops, *d, d, d, d*, are nailed on the upright *a*, at the required distances, according to the space between the

boards on the fence. The bottom boards of the fence
are nailed on first. The bottom block of the board
holder rests upon the bottom board, and is held in position by the brace *c*. The boards can be placed in the holder
like putting up bars, and are guided to their places on
the post by the blocks, *d, d*. The boards can now be
nailed on the posts, and the holding devices moved for
another length. When the boards are too long, they can
be pulled forward a little, and the end sawed, and pushed
back to place. One man using the contrivance, can nail
on nearly as many boards in a day, as two persons with
one to hold the boards in the old way. Figure 37 shows
the manner of using the fence board holders.

REINFORCING A BOARD FENCE.

The old method of topping out a low board fence is
shown at figure 38. Since barbed wire has become

Fig. 38.—STRENGTHENING A BOARD FENCE.

plenty, it is more usual to increase the height of the
fence by stringing one or two strands of that on vertical
slats nailed to the tops of the posts. Yet, in cases where
there are plenty of sound rails left from some old fence,
or plenty of straight saplings, the old method is still a
very cheap and convenient one.

CHAPTER IV.

PICKET FENCES.

A GOOD GARDEN FENCE.

The engraving, figure 39, represents a good, substantial garden fence, that, while somewhat more serviceable than the ordinary kind, may be constructed at less cost. It does not materially differ from the common picket

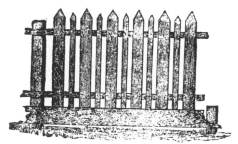

Fig. 39.—A LATH AND PICKET FENCE.

fence, further than that the pickets are put five inches apart, with strips of lath nailed between. The pickets give the necessary strength, while the lath, as a shield against poultry, or rabbits and other vermin, is equally as good at one-sixth the cost. An old picket fence surrounding a garden or yard, may be "lathed" in the manner here indicated at little expense.

A SOUTHERN PICKET FENCE.

The picket fence in very general use in the Southern States, is shown in figure 40. It will be observed that the pickets, instead of terminating in an equal-sided point, have but one slanting side, while the other is straight. Such a fence looks quite as well as one with the other style of points, and is exceedingly neat and

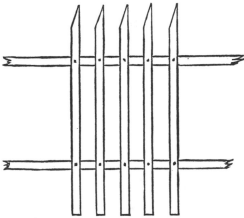

Fig. 40.—SOUTHERN PICKET FENCE.

serviceable along the line of the street, or to mark the boundary between two estates. To facilitate the sawing of the pickets, the bench or horse represented in figure

Fig. 41.—BENCH FOR SAWING PICKETS.

41 is employed. This has a stop at one end, while near the other end are two upright pieces to serve as guides in sawing. The edge of one of these is far enough in the rear of the other to give the desired slope. In saw-

ing, the saw rests against these guides, as shown by the dotted lines. In a picket fence, the point where decay commences, is where the pickets cross the string pieces. Water enters between the two, and decay takes place which is unsuspected until the breaking of a picket reveals the state of affairs. The string pieces and the pickets, at least upon one side, should be painted before putting them together, and nailed while the paint is fresh.

FENCES OF SPLIT PICKETS.

In localities where sawed timber is expensive, and split timber is readily obtained, a very neat picket fence may

Fig. 42.—A FENCE OF SPLIT STUFF.

be made with very little outlay, by using round posts, split stringers, and rived pickets, as shown in the engraving, figure 42. The stringers are eight to twelve feet in length, and usually one of the flat sides is sufficiently smooth for receiving the pickets. Let the stringers project a few inches beyond each post, adding strength to the fence, and should the posts decay, new ones may be driven in on either side, and the stringers readily attached by heavy nails or spikes. With timber that splits freely, a man can rive out five or six hundred pickets in a day. The construction of the fence is plainly shown in the above engraving.

Figure 43 represents a fence made entirely of split timber, the only cash outlay being for nails. This may be made so as to turn, not only all kinds of stock, but rabbits, etc. The pickets are sharpened, and driven six

Fig. 43.—CHEAP FENCE OF SPLIT TIMBER.

or eight inches into the ground, and firmly nailed to a strong string-piece at top.

Another good substantial fence is represented by figure 44, which, though somewhat expensive, is especially

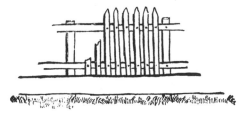

Fig. 44.—COMMON PICKET FENCE.

adapted for yard, orchard and vineyard enclosure. This needs no explanation. The posts should not be set further than eight feet apart; two by four inch scantlings should be used to nail to, and split palings should be nailed on with annealed steel nails.

———◆◇◆———

ORNAMENTAL PICKET FENCES.

The fence shown in figure 45 may be constructed with flat pickets, three inches wide and three feet five inches long. The notches in the pickets are easily made with a compass saw, or a foot-power scroll-saw. The top and bottom pieces between the pickets may be

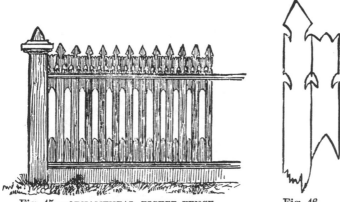

Fig. 45.—ORNAMENTAL PICKET FENCE.   Fig. 46.

painted some other color than the fence, if so desired. Any carpenter should be able to construct it at a small advance over a fence made from plain pickets, making the pattern as in figure 46.

A plainer, but still very neat form of picket fence is

Fig. 47.—A PLAINER PICKET FENCE.

illustrated at figure 47. The intermediate pieces are notched at one end and square at the other.

RUSTIC PICKET FENCES.

When the farmers on the prairies prevent the spreading of the prairie fires, young oak and hickory saplings spring up as if by magic near all the wooded streams. These saplings come from huge roots whose tops have

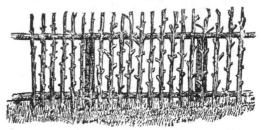

Fig. 48.—RUSTIC SAPLING FENCE.

yearly been destroyed by fire. In that section farmers often construct a very neat rustic fence from two or three year old saplings, having the appearance of figure 48. The rustic pickets are trimmed so as to leave the branches projecting about two inches, and are nailed on with four-penny nails. A fence of this kind would not last long, unless the pickets, posts, and rails were free of bark, or saturated with crude petroleum.

A very neat and picturesque fence for a garden or

Fig. 49.—RUSTIC PICKET FENCE.

a lawn is shown at figure 49. It is made of round poles, with the bark on, the posts being of similar material. Three horizontal bars are nailed to the posts at equal intervals, the slats or pickets woven into them and then nailed in place. One or two coats of crude petroleum, applied to this and other rustic work at first, and renewed every year, adds to its appearance and greatly increases its durability.

LIGHT PICKET FENCES.

For enclosing poultry yards, garden and grounds, a cheap fence with pickets of lath often serves a good purpose. If not very durable, the cost of repair or renewal is light. Figure 50 shows one of this kind, which is sufficiently high for the Asiatic and other heavy and quiet fowls. The panels are sixteen feet long, and are

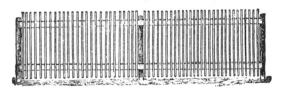

Fig. 50.—PANEL OF PICKET FENCE.

composed of two pieces of ordinary six-inch fencing, for top and bottom rails, with lath nailed across two and a half inches apart; the top ends of the lath extending ten inches above the upper edge of the top rail. Posts, three or four inches through at the top end, are large enough, and, after sharpening well, can be driven into the ground by first thrusting a crow-bar down and wrenching it back and forth. A post is necessary at the middle of each panel. Both rails of the panel should be well nailed to the posts. These panels may be neatly and rapidly made in a frame, constructed for that purpose. This frame, shown in figure 51, consists simply of three cross-pieces of six by six, four feet long, upon which are spiked two planks one foot wide and three feet apart, from outside to outside. Four inches from the inner edge of each plank is nailed a straight strip of inch stuff, to keep the rails of the panel in place while the

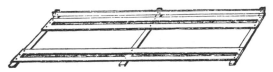

Fig. 51.—FRAME FOR MAKING FENCE.

lath are being nailed on. Against the projecting ends of the cross-pieces, spike two by six posts twelve inches long ; on the inside of these posts nail a piece of six-inch fencing, to serve as a stop, for the top ends of the laths to touch, when nailing them to the rails. These panels can be made in the shop or on the barn floor at odd times, and piled away for future use. Nail a wide bottom board around on the inside of the enclosure after the fence is in position.

Figures 52 and 53 show lath fences high enough for all kinds of poultry. The posts in figure 52 are eight feet apart. A horizontal bar is nailed to the posts six

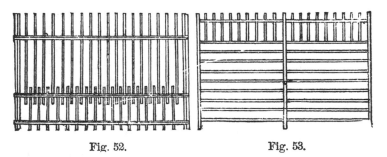

Fig. 52.   Fig. 53.

inches above the ground, a second one eighteen inches, and a third four and a half feet. To two lower strips nail laths that have been cut to half length, first driving the lower part of the laths two inches into the ground.

One advantage of this fence is, that the two strips near the bottom, being so close together, sustain pressure from dogs or outside intruders better than any other fence constructed of lath, and dispenses with a foot-wide board, so generally used.

The cheapest lath fence is made with the posts four feet apart, first sawing them in two lengthwise at a sawmill, and nailing the lath directly to the posts without the use of strips. The two upper laths have short vertical pieces fastened to them with cleat nails, and present points to prevent fowls alighting on the fence. Such a fence (figure 53) will cost, for four feet, one-half post, three cents; twenty laths, eight cents; and the nails, three cents, per running foot, six feet high, or one-half cent per square foot.

HAND-MADE WIRE AND PICKET FENCES.

A very desirable and popular fence is made of pickets or slats woven into horizontal strands of plain wire. Sev-

Fig. 54.—SIDE VIEW OF BENCH.

eral machines have been invented and patented for doing this work, but it can be done by hand with the aid of the bench illustrated herewith. The wire should be a little larger than that used on harvesting machines, and annealed like it. The bench, of which figure 54 is a side view, and figure 55 a top view, should be about sixteen feet long and have a screw at each corner for raising and lowering the holding bars. For the screws at the ends of the frame one-half to three-fourth-inch iron rod will answer. The wire is twisted close and tight to the slats, and given two or three twists between them. If the

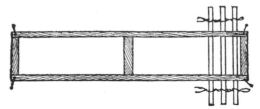

Fig. 55.—TOP VIEW OF BENCH.

slats are of green stuff, fasten the wire to them with small staples, to prevent their slipping when they shrink. The fence is fastened to the post with common fence staples.

Fig. 56.—PORTION OF THE FENCE.

When this style of fence is used on one side of a pasture or highway, its effectiveness may be increased by a single strand of barbed wire stapled to the posts above the pickets, and a strand of plain wire strung along the bottom to stiffen it. The fence will then be as in figure 56. Such a fence will last many years, and for most sections of the country is the best and cheapest combined cattle and hog fence that can be made. For a garden fence it is equal to the best picket, and at one third of the cost. By having the slats sawed about one-half-inch thick, two inches wide, and five to six feet long, it makes an excellent fence for a chicken yard, as it can be readily taken down, moved, and put up again without injuring it in the least. For situations where appearances are secondary importance, round slats are equally as good as pickets. A farmer in Wisconsin planted a few white willow trees the year that he made some fences of this kind. When the fence began to need repairs, the willows had attained such a growth that their trimmings furnished all the material needed then and each year thereafter.

FENCE OF WIRE AND PICKETS.

The fence shown in figure 57 has been introduced in some sections, and is becoming more popular every

Fig. 57.—FENCE OF WIRE AND PICKETS.

year. The posts are set ten feet apart, and are so placed that they will come on the right and left side of the fence, alternately. The pickets are split from oak, or any other hard wood, and are four or five feet long, and an inch and a half or two inches wide. When the posts are set, brace the one at the end of the line, and

fasten the ends of two number nine, unannealed wires
to it. Stretch the wires along to the other end of the
line, and a few feet beyond the last post. One pair is
to be stretched near the top of the posts and one near the
ground. When the wires are stretched taut, fasten them
to some posts or other weight that will drag on the
ground ; the upper and lower wires should be fastened to
separate weights, and these should be heavy enough to
keep the wires at a great tension. Having done this, you
are ready to commence building the fence. One man
spreads the strands, while another places the picket between them; the other end of the picket is then raised
up and placed between the upper wires, and then driven
up with an axe or mallet. In inserting the pickets, the
wires are to be crossed alternately, as shown in the engraving. The pickets should be dry and should be about
three inches apart. It takes two persons to build this
fence successfully, but it can be built more rapidly by
three; one to spread the wires, one to place the picket in
position, and one to drive it home. This is especially
adapted for a line or other fence which is not required
to be often moved. It is fastened to the post by nailing
one of the pickets to it with common fencing nails.
Fences of this kind are also made with straight, round
limbs of willow or other trees in place of the split pickets.
Several different machines have been patented for making this style of fence.

CHAPTER V.

BARB-WIRE FENCE.

The invention of barb wire was the most important event in the solution of the fence problem. The question of providing fencing material had become serious, even in the timbered portions of the country, while the great prairie region was almost wholly without resource, save the slow and expensive process of hedging. At this juncture came barb wire, which was at once seen to make a cheap, effective, and durable fence, rapidly built and easily moved. The original patent for barb wire was taken out in 1868, but it was not until six years later that an attempt was made to introduce it into general use, and more than ten years elapsed before the industry attained any considerable magnitude. The rapidity and extent of its subsequent growth will be seen by the following table, showing the estimated amount of barb wire manufactured and in use during the years named, the estimated length being in miles of single strand:

YEAR.	TONS.	MILES.	YEAR.	TONS.	MILES.
1874	5	10	1881	60,000	120,000
1875	300	600	1882	80,000	160,000
1876	1,500	3,000	1883	100,000	200,000
1877	7,000	14,000	1884	125,000	250,000
1878	13,000	26,000	1885	130,000	260,000
1879	25,000	50,000	1886	135,000	270,000
1880	40,000	80,000			
		TOTALS		716,805	1,433,610

There are now fifty establishments engaged in the manufacture, and the output for 1887 is estimated at 140,000 tons.

Barb wire is not without its drawbacks as a fencing material, the most common one being the liability of seri-

(43)

ous injury to valuable domestic animals coming in contact with the sharp barbs. Many means have been devised for overcoming this evil. Some of them are illustrated in the next chapter. The direct advantages

Fig. 58.—THE KELLY BARB WIRE.

of barb wire are: First—economy, not only in the comparative cheapness of its first cost, but also in the small amount of land covered by it. Second—effectiveness as a barrier against all kinds of stock, and a protection against dogs and wild beasts. Third—rapidity of construction and ease of moving. Fourth—freedom from harboring weeds, and creating snow drifts. Fifth—durability.

Barb wire, like the harvester, the sowing machine, and

Fig. 59.—HORSE-NAIL BARB.

most other valuable inventions, has attained its present form from very crude beginnings. The original barb wire consisted of double-pointed metallic discs, strung loosely upon plain wire. The next step was to twist this with another wire, as shown in figure 58.

Another crude beginning was the "horse-nail barb,"

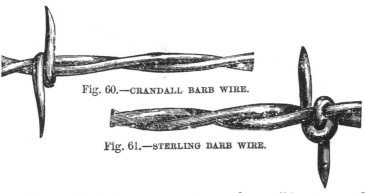

Fig. 60.—CRANDALL BARB WIRE.

Fig. 61.—STERLING BARB WIRE.

which consisted of a common horse-shoe nail bent around a plain wire, and the whole wrapped spirally with a smaller wire, as shown in figure 59. Various forms of two-pointed and four-pointed barb wire are manufactured, the principal difference being the shape of the barbs and

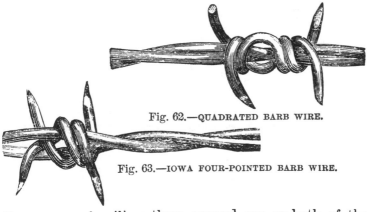

Fig. 62.—QUADRATED BARB WIRE.

Fig. 63.—IOWA FOUR-POINTED BARB WIRE.

the manner of coiling them around one or both of the strands. A few of the leading styles are illustrated herewith. Figures 60 and 61 show two varieties of two-pointed barb wire.

Of the numerous styles of four-pointed wire, three typical forms are illustrated in figures 62, 63, and 64. The Glidden patent steel barb wire is made in three

Fig. 64.—LYMAN BARB WIRE.

styles, as shown in figures 65, 66, and 67.   Figure 65 shows the two-point wire, in which, like the others, the barb is twisted around only one of the wires.   Figure

Fig. 65.—GLIDDEN PATENT STEEL TWO-POINT.

66 shows the "thick-set" which has barbs like the other, but set closer together for such purposes as sheep folds, gardens, or other places, which require extra protection.

Fig. 66.—GLIDDEN PATENT STEEL " THICK SET."

The four-point barb wire, figure 67, has barbs of the same form as the two other styles, that is a sharply pricking barb attached to one of the wires of the fence strand, upon which the other wire is twisted, holding the barb firmly in place.   The barb is at right angles to the wire, and does not form a hook, but a straight short steel thorn. A sharp point which inflict an instantaneous prick repels an animal more safely than a longer and duller barb.

Barb wire of nearly, if not quite all the popular kinds, is shipped from the factory on strong spools, each holding

Fig. 67.—GLIDDEN PATENT FOUR-POINT.

one hundred pounds in weight, or eighty rods in length. These spools are bored through the center to admit a stick or bar, which can be used as an axle in unreeling the wire.   The following table shows the weight of wire required for fencing the respective areas named :

AREA.	LENGTH OF BOUNDARY.	WEIGHT OF WIRE.	
		1 Strand. Lbs.	3 Strand. Lbs.
1 Acre	60 Rods.	67	202
5 Acres	3/8 Mile.	167	400
10 Acres	1/2 Mile.	183	548
20 Acres	3/4 Mile.	273	820
40 Acres	1 Mile.	365	1095
80 Acres	1 1/2 Mile.	547	1642
160 Acres	2 Miles.	730	2190

It will be observed that the larger the area enclosed, the smaller is the amount of fence required per acre.   The cost of fence complete can be estimated by adding to the amount of wire indicated in the last column, the cost of sixty posts, and three and three quarter pounds of staples, for every sixty rods. To ascertain the weight of wire required for any desired number of strands, multiply the

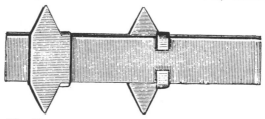

Fig. 68.—BRINKERHOFF STEEL STRAP AND BARB.

figures of the first column of " weight of wire " by the number of strands proposed to be used.

There is a kind of barb fencing in which flat steel straps are employed instead of wire. In the form shown in figure 68, the barbs are bent around a plain strap and the whole is then galvanized, which firmly fixes the barb.

Fig. 69.—ALLIS PATENT BARB.

Another form shown at figure 69 consists of a solid piece of steel, ribbed through the middle, and with barbs cut on both edges. These and similar forms are more expensive than wire, and are employed only in limited quan-

Fig. 70.—BRINKERHOFF FENCING TWISTED.

tities for enclosing lawns, paddocks, etc. Still another form is like that shown in figure 70, without barbs, and twisted. This is much used to enclose lawns and ornamental grounds. It is light, neat and strong, does

not harbor weeds or make snow drifts, but is compara-
tively expensive, as five or six strands are required to
make an effective fence.

Still another form of unarmed fencing is shown in
figure 71. It is simply the ordinary wire without barbs,

Fig. 71.—TWO STRAND TWISTED WIRE FENCING.

and is used in limited quantities for fencing ornamental
grounds, barnyards, etc.

STEEL FENCE STAPLES.

For fastening barb wires to the post nothing has been
found so satisfactory as staples made for the purpose
from No. 9 steel wire. They are cut with sharp points

Fig. 72.—1¼-INCH STAPLE. Fig. 73.—1¾-INCH STAPLE.

Fig. 74.—SQUARE TOP STAPLE FOR BRINKERHOFF FENCING.

to drive easily into the posts, and are of different lengths,
from one inch and a quarter to one and three-quarters.
Figures 72 and 73 show the usual staples for wire, and
figure 74 a staple made specially for strap fencing.

HOW TO SET BARB WIRE FENCE.

The timber for posts should be cut when the sap is dormant. Midwinter or August is a good time to cut post timber. They should be split and the bark taken off as soon as possible after cutting the timber. For end posts, select some of the best trees, about sixteen inches in diameter, from which take cuts eight and a half feet in length, splitting them in quarters for brace posts. They should be set three feet in the ground, which is easily done with a post-hole digger. When setting the brace posts, take a stone eighteen inches to two feet long, twelve inches wide, and six inches thick, which is put down against the post edgewise, on the opposite

Fig. 75.—WELL-BRACED BARB-WIRE FENCE.

side to the brace, as seen in figure 75, putting it down about even with the surface of the ground. This holds the post solid against the brace. A heart-rail, ten feet in length makes a good brace. Put one of the long posts every sixteen or twenty rods along the line of fence, as they help to strengthen it, and set lighter and shorter posts along the line about sixteen feet apart. After the posts are set, two or three furrows should be turned against them on each side, as it helps to keep stock from the wire. Such a fence should be built of a good height. It is better to buy an extra wire than have stock injured There is no pulling over end posts or sagging wire.

To make an extra solid wire fence, brace the posts, as shown in figure 76, on both sides, in order to resist the tension in either direction. Every eighth post should be thus braced, and it makes a mark for measuring the length of the fence, for eight posts set one rod apart, make eight rods, or a fortieth of a mile for each braced post. The braces are notched into the top of

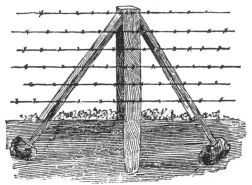

Fig. 76.—A WIRE FENCE WELL BRACED.

the posts, just below the top wire, and a spike is driven through both the brace and the post. The braces abut upon large stones which give them great firmness.

———◦◦———

UNREELING AND STRETCHING BARB WIRE.

The general introduction of barb wire fencing has brought out a great variety of devices for handling the wire. One of these is shown in the illustrations. Two pieces of scantling are attached to the rear end of a wagon from which the box has been removed, as shown in figure 77. A slot near the end of each admits the round stick thrust through the reel of barb wire, to serve as an axle. The end of the barb wire is fastened to the fence post, the team in front of the wagon started up, and

some three yards of wire unreeled.   Then the hind axle
of the wagon is made fast by a chain or rope to the nearest fence post, the hind wheel nearest the fence lifted
from the ground and held there by a wagon-jack or piece

Fig. 77.—DEVICE FOR UNROLLING WIRE.

of board.   One turn is then made in the barb wire, as
shown at *A*, figure 78, to which is attached one end of a
piece of smooth wire, some ten feet long.   The other
end is placed between two screws, *b b*, in the end of the

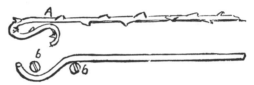

Fig. 78.—FASTENING THE WIRE.

hub, as shown in the illustration.   The wire thus fastened is coiled around the hub, and the operator can
tighten it and the barb wire to which it is attached, by
employing the leverage of the spokes and felloes.

A lighter form of reel holder is shown at figure 79.   It

Fig. 79.—A SULKY WIRE-HOLDER.

is made of two pieces of two by four scantlings fastened
to the axle of a sulky corn plow.   They must be placed

far enough apart to allow the reel or spool to run between them. Make a square axle, figure 80, of some hard tough wood, rounding it where it runs in the slots of the

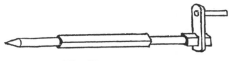

Fig. 80.—THE AXLE.

scantling ; drive it through the hole in the spool, and attach the crank. In moving fence, place the spool on the frame ; remove one end of the wire from the post, fasten it to the spool, and while one man holds the pole and steers and steadies the sulky—he will have to pull back a little—another turns the spool and winds up the wire. When a corner is reached, the wire is loosened, the sulky turned, and the winding continued. When the end of the wire is reached, it is carefully loosened from the post, and firmly fastened to the spool.

It is best to have a separate spool for each wire, especially if they are of great length. The same contrivance may be used for unreeling the wire. Attach a gentle horse to the sulky, fasten the pole securely to the hames,

Fig. 81.—A SLED WIRE-HOLDER.

and have a boy lead him slowly along the fence line. once in fifty yards stop the horse, grasp the handle, move forward very slowly, and draw the wire straight and taut. If no sulky plow is at hand, a light " double-ended" sled, shown in figure 81, may be used. A man holds the short pole extending from one end, steadying and pushing a little, while the other winds the reel. The sled is drawn forward by the wire as it is wound on the reel. To unreel, attach a slow horse to a chain or

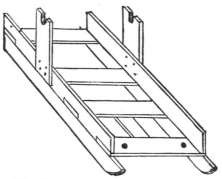

Fig. 82.—ANOTHER SLED FOR WIRE.

rope fastened to the opposite end of the sled. A man must walk behind the horse and hold the pole to steady the sled. Managed in this way, the removal of a barbed wire fence is not at all the formidable operation that has been supposed ; it can be taken down and set up again, easily, safely, and quite rapidly. Figure 82 shows an-

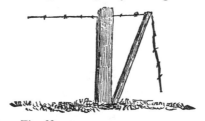

Fig. 83.—TIGHTENING THE WIRE.

other form of home-made sled, which is very useful for carrying rolls of wire for making a fence. The roll is supported on a rod, which has round ends to fit into the uprights, and which turns in the slots. When the wire is run out, the end is fastened to the clevis on the centre beam, and a notched stake, figure 83, being put under the wire, the sled is drawn up to tighten the wire, which is then stapled. This sled is useful for many other purposes, and is large enough to carry five rolls of the wire, so that by going back and forth, the whole of the fence can be put up very quickly. It is drawn by one horse, the draft chain being fastened to the front beam.

WIRE STRETCHERS.

For stretching barb wire there are various implements in the market, and other quite simple and effective devices can be made on the farm. Figure 84 shows the

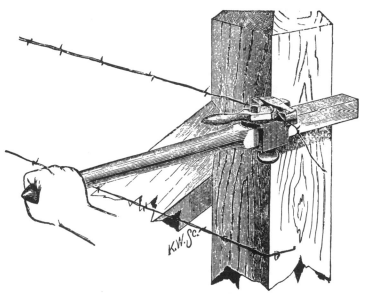

Fig. 84.—THE CLARK STRETCHER.

Clark stretcher and the manner of using it. Another stretcher, called the "Come Along" stretcher, figure 85, is used not only for tightening the wires, but also for handling it, in building or moving fences.

The useful wire stretcher, figure 86, consists of a
mowing machine knife-guard, bolted to a stout stick ;

Fig. 85.—THE "COME ALONG" STRETCHER.

one curved, as shown in the lower engraving, is preferable to a straight one, as it will not turn in the hand.
When using it, the wire is held firmly in the slot, and
may be easily stretched by applying the stick as a lever.

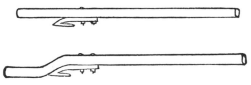

Fig. 86.—HOME-MADE WIRE STRETCHERS.

Another kind of a wire-stretcher may be made of hard
wood or of iron or steel bars.   It consists of three pieces,
two arms and a splicer, fastened together in the manner
shown in figure 87, leaving a slot near one end to
hold the wire.   The longer arm is made immovable upon
the splice by means of two or more heavy bolts, while the

shorter arm is pivoted by one bolt. This allows the slot to be opened to receive the wire. The short arm is sharpened so that it may be stuck into a post, or the side of a building, if convenient. By placing this lever behind a post, one man can stretch thoroughly a long string of wire. When one man is doing the work alone, he can stretch the wire, fasten the lever back by means of a stick

Fig. 87.

driven into the ground before it, and then go back and drive the staples. The short end of the lever should be about twelve inches long, and the long arm three or four feet, or even longer.

The stretcher shown in figure 88 is made of hard tough wood or iron. The wire is passed through the slot, the barbs preventing it from slipping. The arm at right angles to the lever is used to measure the distance of the strands. When the lever is set against the post, the arm

Fig. 88.—STRETCHER AND GAUGE.

rests on the strand below. By sliding it up or down, the distance between the strands is regulated.

Figure 89 shows another stretcher, that can be made by any blacksmith. The toothed cam holds the wire so that it will not slip. A block and tackle are often found useful to draw the wires with. The rolls of wire are paid out of a wagon body, and when the wire is to be drawn up, the grip is put on at any point, the tackle is attached, and one horse draws it as tight as it needs be.

A wire fence needs frequent drawing up or it sags and

becomes useless. The alternate contraction and expansion caused by change of temperature soon stretch the wire, to say nothing of other causes. The cheap and ef-

Fig. 89.—GRIP FOR FENCE WIRE.

fective method employed by telegraph companies is illustrated in figure 90. It consists of a pair of grip tongs and a set of small tackle-blocks. The tongs may be made by any blacksmith, and the blocks are sold at all hardware and tool stores. An iron hook is used to cou-

Fig. 90.

ple the tongs to the block, and as the wire is drawn up, the free end of the rope may be given a turn around the same post, to hold it while the staple is tightened to hold the wire.

———◦◦◦———

SPLICING BARB WIRES.

The accompanying engravings show an iron implement for splicing wire and the manner of using it. To make this splicer take a bar of half inch round iron, nine inches long. Heat about three inches of one end and hammer it flat until it is one inch wide. With a cold chisel cut a one-fourth inch slot a quarter of an inch from the right side and an inch deep, as seen in figure

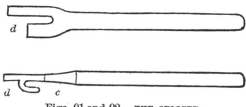

Figs. 91 and 92.—THE SPLICER.

91. Bend the part marked *d*, so that it will be a quarter inch from the flat part, as shown in figure 92. The lower part of the slot *c* should be about a half inch from the bend at *d*. Smooth with a file. To use it let *e* and

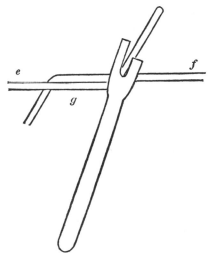

Fig. 93.—MAKING THE SPLICE.

f, figure 93 represent two wires to be joined. Bend the ends so they are nearly at right angles. Hold them with pincers at *g ;* place the hook of the splicer on the wire *f*, while the wire *e* falls into the slot. Twist the pieces around the wire *f*, when one half of the splice is

made. Repeat the operation for the other end. Use
about four or five inches of each wire to twist around
the other. Another form of splicer, shown in figure 94,
is made of cast iron, and is used in the same manner as

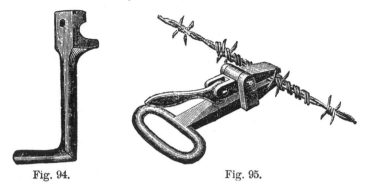

Fig. 94.　　　　　　　　Fig. 95.

the first. Figure 95 shows the manner of holding the
wire with nippers made for the purpose, and the finished splice.

BUILDING WIRE FENCES ON UNEVEN GROUND.

One of the great perplexities about building wire
fences on rolling ground, is how to make the posts in
the hollows remain firm, for the pull of the wire in wet

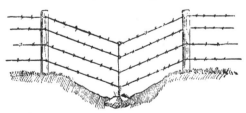

Fig. 96.—FENCE ON UNEVEN GROUND.

weather, or when the frost is coming out, lifts them and
causes the wire to sag, and they cease to be an effective
barrier. Posts should not be used in the lowest depres-

sions, but in their place at the lowest spots a heavy stone should be partially sunk into the ground, about which a smooth fence wire has been wrapped, as seen in figure 96. When the fence is built, the fence wires are brought down to their place and the wire about the stone is twisted first about the lower wire, then the next, and so on to the top. This prevents the wire from raising, and does away with all trouble of the posts being pulled out by the wires. In fencing across small streams the same plan is successful.

CHAPTER VI.

FENCES OF BARB WIRE AND BOARDS.

COMBINED WIRE AND BOARD FENCE.

A very cheap fence is made of two boards below and three strands of barb wire. To make the fence pig-proof without the boards, five strands of wire, three inches

Fig. 97.—MANNER OF BRACING END-POST.

apart, would be required at the bottom. Two common fencing boards will occupy the same space, when placed three inches apart, and cost less. But for the upper part of the fence, wire is much cheaper than boards. The most considerable item in this greater economy is the saving of posts. The wire requires a post every sixteen feet ; hence half the posts are saved. A stout stake, driven midway between the posts, holds the center of the boards in place. These stakes need extend only eighteen inches above ground. Posts that have rotted off in the ground will be long enough for these stakes. Some say that the posts can be set thirty feet apart, but sixteen feet is better. The posts should be at least thirty inches in the ground and well tamped. It is easy to stretch the wire. Its durability depends upon the quality of the wire and posts, and the proper setting of them. Nail on the two boards, three inches apart ; the first strand is six inches above the top board, the second strand is twelve inches above the first, and the third sixteen inches above the second. When banked up, as hereafter described, this fence will turn all farm stock. An im-

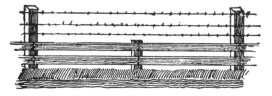

Fig. 98.—SECTION OF FENCE COMPLETED.

portant point is the bracing of the end-posts. If this be neglected or improperly done, the fence will be a failure. Figure 97 shows how the end-post should be braced. It should be a large post and set at least three feet in the ground. The short post which holds the lower end of the brace, should also be well set. Wrap the wire around the end-post several times, and drive staples to hold it on all sides. If the line of fence is more than forty rods long, at least two posts at each end should be braced. After the posts are set, and before attaching the boards or wire, plow a deep furrow along each side, throwing the earth inward. This makes a bank along the line, allowing the fence to be several inches higher ; and the furrow drains the water away from the posts, and also restrains an animal that may be tempted to jump the fence. A section of the completed fence is shown in figure 98. Do not hang pieces of tin, etc., upon the top strands of wire, as often recommended, that the animals may see the fence, and be able to avoid it, because it is never necessary.

A modification of this combined fence is shown in figure 99. It is made of one rail along the top, and three wires below. After setting the posts plow a fur

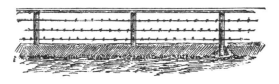

Fig. 99.—A CHEAP AND GOOD FENCE.

row two feet from the posts on each side, throwing the furrow slice towards the fence, and forming up the ridge neatly with a spade ; then stretch the three wires, and nail a two by four scantling edgewise. To prevent an unpleasant sagging of the rails, the posts should be eight feet apart, and the rails sixteen feet long. For common fencing, good straight poles will answer well.

A BRACKETED FENCE.

The features shown in figure 100 are : first, in having two six-inch boards at the bottom. Second, in placing the wires very close together. It being necessary to have barbs only on one side of each space between the wires, plain galvanised wire may be used for every alternate strand, thus greatly lessening the expense. Third, by the use of strips and short stakes, the posts may be placed sixteen feet apart, and the fence remain as perfect as if there were posts every eight feet. Fourth, to make the fence man-proof, make use of a bracket of three-eighth-inch iron, or of one by two-inch wooden strips. The form of the brackets is shown in figures 101, 102 and 103. A barb-wire is attached to the short arm of the brackets, which are fastened to the posts in such a manner as to

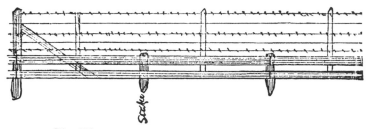

Fig. 100.—ONE PANEL OF IMPROVED WIRE FENCE.

stretch two wires on the same horizontal plane, and fifteen inches apart. The material required for each panel of the fence shown in figure 100, are: Two posts, three barb-wires, two plain wires of No. 12 galvanized iron, two six-inch boards, sixteen feet long, three stakes about three feet long, and sharpened at one end, four strips, four feet long and one and one-half-inch square. To build the fence: Lay off the ground by setting small pegs eight feet apart, then dig the holes, and set the posts at

Fig. 101.—IRON BRACKET.

every fourth peg. Drive the sharpened stakes into the ground at the three pegs between the posts, so that the top of the stakes will be nineteen inches above the ground. Nail the boards on the first stake near the

ground, and the second one three inches above the first.
Then mark off the place for each wire on the first post,
fasten the bottom wire, and put up as far as the first
stretching post; then add the other wires, using first a
barb-wire, and then a smooth one. The wires should be
fastened to the posts with long staples. The strips are
to go in the middle of the eight foot spaces; they should
not quite touch the ground; fasten them to the boards

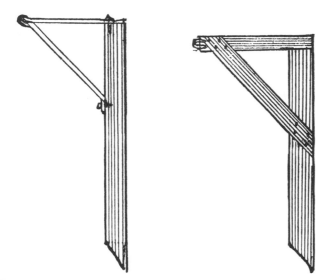

Fig. 102.—ATTACHED BRACKET.   Fig. 103.—WOODEN BRACKET.

with nails and to the wire with short staples. These
strips can be made of poles or saplings, and the stakes of
short or crooked pieces from the posts. To attach the
man-proof part: If the brackets are of wood, nail them
to the posts, sawing off the horizontal arm to fifteen
inches from the top wire, as in figure 103 ; stretch the
wire and fasten to the end. If the brackets are of iron
figure 102, spike the horizontal arm to the top of the
post, then put up the barb-wire loose under the oblique
arm, and stretch it. Then spike the foot of the oblique

arm to the post, and slip the wire into the angle, and close the bracket by closing the arms on the wire. Figure 102 shows the method of attaching the iron bracket to the post.

DOG-PROOF FENCES.

Figure 104 shows a sheep-yard fence, built of wire and boards, as a safeguard against vicious dogs.   It consists of

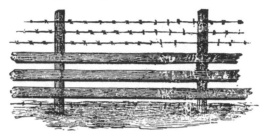

Fig. 104.—A FENCE AGAINST DOGS.

ordinary posts, and three lengths of boards, with an equal number of barb-wires for the upper portion, and a single strand placed near the ground.   The sheep are in no danger of injuring themselves with such a fence, and it is an effective barrier to blood-thirsty dogs.

Figure 105 shows a cheaper fence for the same purpose.   It has one strand of barb wire below the boards,

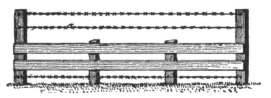

Fig. 105.—A CHEAPER FENCE.

which prevents attempts of dogs to dig under it. For fencing sheep against dogs, the " thick-set " barb wire is the most effective of any.

CHAPTER VII.

HEDGES.

THE BEST HEDGE PLANTS.

The first emigrants from England to the American shores brought with them memories of green hedge-rows, like those which still adorn the motherland. But they found the country whither they had come covered with a dense growth of timber, which furnished abundant material for fences. Hedges were almost unknown in this country until after civilization had reached the treeless prairies. Then, the want of fencing material turned attention to hedges, and they became so popular that many miles of them were planted, not only in the prairie region, but also in the more eastern States, where cheaper fencing material was plenty. Now the invention of barbed wire supplies a material so cheap and easily put in place, that hedges have ceased to be regarded as economical for general farm purposes. But they have by no means gone wholly out of use. As a boundary fence, especially upon the roadside, there is much to be said in favor of the hedge. Nothing gives a neighborhood such a finished rural aspect, as to have the roads bordered by hedges. The grounds around the summer cottages on the New Jersey coast, and other popular summer resorts, are largely enclosed with hedges. For interior divisions, as they cannot be removed, they are not to be commended. An orchard, the most permanent of all the plantations upon the farm, may be appropriately enclosed by a live fence. Hedges are either protective barriers, really live fences, or merely ornamental. In properly regulated communities, where cattle are not al-

(67)

lowed to run at large, the roadside hedge may be ornamental, while one around an orchard should be able to
keep out animals and other intruders. After many ex-
periments and failures, the Osage Orange (*Maclura
aurantiaca*), has been found to make the best hedges.
Being a native of Arkansas, it has been found to be hardy
much farther North, and may be regarded as the most
useful hedge plant in all localities where the winter is
not severe. Where the Osage Orange is not hardy,
Buckthorn, Japan Quince and Honey Locust are the best
substitutes. Honey Locust is a most useful hedge-plant,
as it is readily raised from seed, grows rapidly, bears
cutting well, and in a few years will make a barrier that
will turn the most violent animal.

PLANTING AND CARE OF OSAGE HEDGES.

The first requisite for a hedge of any kind is to secure
thrifty plants of uniform size. Osage Orange plants are
raised from seeds by nurserymen, and when of the right

Fig. 106.—BADLY PLOWED GROUND.

size, should be taken up in autumn and "heeled in."
The ground, which it is proposed to occupy by the
hedge, should be broken up in autumn and then re-
plowed in spring, unless it is a raw prairie sod, which
should be broken a year before the hedge is planted. It
is a very usual, but very bad practice, to plow a ridge
with a back-furrow, as shown in figure 106. This leaves
an unplowed strip of hard soil directly under the line
upon which the hedge is to stand. When harrowed, it
appears very fair on the surface, but it is useless to ex-

pect young plants to thrive on such a bed of hard soil, and its result will be as seen in figure 107. The first growth is feeble, irregular, and many vacant spots ap-

Fig. 107.—HEDGE PLANT ON HARD RIDGE.

pear. The land should be plowed as in figure 108. When the sod is rotted, the land should be harrowed lengthwise of the furrows, and the dead furrow left in the first

Fig. 108.—PROPERLY PLOWED GROUND.

plowing closed by twice turning back the ridge. There is then a deep, mellow, well-drained bed for the plants in which the roots have room to grow and gather ample nutrition. Figure 109 shows the effect of this kind of

Fig. 109.—HEDGE PLANT IN MELLOW SOIL.

cultivation. As a barrier against stock, or a windbreak, it is best to plant in double rows, each row being set oppo-site the spaces in the other, thus: *   *   *   *   *
                                                                                                  *   *   *   *
It is highly desirable that the hedge should be in true, uniform rows, either straight or in regular curves. This can be done only by setting closely to a line. Osage Orange plants may be raised from seed, but as this is a

difficult operation, it is usually best to buy young plants
from a reliable nurseryman.   They are best cut down to
about six inches high, and the roots partially trimmed.
It is an advantage to "puddle" the roots, which is done
by dipping them in a mixture composed of one-half
earth and half fresh manure from the cow stable, wet
to the consistency of a thin paste.   There are various
methods of setting the plants.   Some use a trowel with
a blade about ten inches long ; others a dibble, and a
larger number than either of the others, a spade.   For
setting long lines, in situations where appearances are of

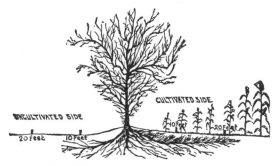

Fig. 110.—EFFECT OF CULTIVATION.

secondary importance, young Osage plants are set very
rapidly by running a furrow where the rows are to stand,
laying the plants with their roots spread on the mellow
soil, one side of the furrow.   A furrow is next turned
upon the roots, and the plants which may have been dis-
arranged are restored by hand.   A tread of the foot will
consolidate the earth around each plant.   Unless the
subsoil is naturally very porous, the ground must be
thoroughly underdrained.   A line of tiles should be laid
six or eight feet from the line of the hedge. The ground
for four or five feet on either side of the hedge, should
be kept thoroughly cultivated the first three or four
years after planting.   This cultivation is to be done
early each season and cease the first of July, to give the

new wood a chance to ripen. The plants should be per-
mitted to grow the first year undisturbed. The following
spring, the hedge should be cut off close to the ground
with a scythe or mowing machine, and all vacancies
where plants have died out or been thrown out by frost,
should be filled. The ground on both sides of the ridge
is to be kept well cultivated. Figure 110 shows the difference in root growth in cultivated and uncultivated
ground.

A thick growth of young shoots will appear, and
these are to be cut back to four inches high, the middle
of summer and again in September. The object is to
obtain a dense growth close to the ground. The third
year the pruning is to be repeated, only the shoots must
be left four to six inches above the last previous cutting.
The lateral shoots which are near the ground, are to be
left undisturbed. The trimming should be such as to
leave the hedge broad at the base, with a regular slope to
the summit like a double-span roof.

Another method is to permit the hedge to grow un-
trimmed for four or five years. It is then plashed, or

Fig. 111.—HEDGE "PLASHED."

laid over sidewise. This is done by cutting the plants
about half through on one side with a sharp axe, and
bending them over as shown in figure 111. The hedge
is first headed back and trimmed up to reduce the top.
In a short time new shoots will spring from the stubs and
stems, making a dense growth of interlacing stems and

branches. Another method of laying a hedge, is to dig away a few inches of earth on one side of each plant to loosen the roots, then lay the plant over to the desired angle and fasten it there. The earth is then replaced around the roots, and tread down firmly. We believe that a patent is claimed for this process, but its validity is seriously questioned.

It is essential that hedges, whether planted for ornament or utility, shall be kept in shape by trimming every year. It is less labor to trim a hedge three times during the year, when the branches are small and soft, than once when the branches have made a full season's growth. If the hedge is trimmed once in June and again in August, it will be kept in good shape, and the labor will be less than if the trimming was put off until spring. In August the branches can be cut with shears or a sharp corn knife. The foliage on them will aid in their burning, when they have dried a few days in the sun. The thorns are not so hard as in the spring. The brush will be less, and on account of their pliability and greater weight, will pack into the heap much better. If trimmed in August, the hedge will not make any considerable growth during the fall. August trimming does not injure the hedge, rather helps it, as it tends to ripen the wood, preventing a late Autumn growth to be injured by the winter. The loss of sap is less than when the trimming is done in the early spring, as then the wounds are larger, and do not heal before the sap flows. Do not neglect to burn the brush as soon as it has dried sufficiently. If allowed to remain on the ground, it will harbor mice and other vermin. Trim the hedge in August and burn the brush. The trimming should be done in such a manner as to expose the greater amount of foliage to the direct action of the light, air, rain and dew. This is attained by keeping the sides at every trimming in the form of sloping walls from the broad base to the summit like a double-span roof. They are sometimes trimmed with vertical sides and broad, flat top, but this is not a favorable plan for permanency. The lower leaves and stems die out, leaving an unsightly open bottom of naked stems, with a broad roof of foliage above. Such trimming and its results have done much to bring hedges into disrepute.

HEDGES FOR THE SOUTH.

The Osage Orange is a native of the Southwestern States, and flourishes on good soil anywhere in the South. Yet there are certain succulent plants which grow so rapidly in the South, and require so little care, that they are very successfully employed for hedges in the Gulf States. One of these if the *Yucca gloriosa,* **or**

Fig. 112.—CACTUS HEDGE.

Spanish Bayonet. Its natural habit of growth is to produce a dense mass of leaves on a long stem. But by cutting back the growth of the stiff, armed leaves is produced low down, and a hedge of this soon becomes an impassable barrier. Large panicles of beautiful white blossoms are produced at the summit, making such a hedge very ornamental during the flowering season. Various species of cactus are also employed in the South-west for hedges. In some of the Middle-Western States may be seen a hedge like figure 112. At some distance from the highway, a field had been enclosed with the tree cactus, which there grows only from four to ten **feet** high. The plants that were in the line of the fence were left growing, and those cleared from the field were woven into a formidable barrier to anything larger than a rabbit. While no two rods in this fence are alike, its general appearance is like that shown in the engraving.

———◆◇◆———

ORNAMENTAL HEDGES AND SCREENS.

Hedges and screens for ornamental purposes alone, do not come strictly within the scope of this work, but we will briefly mention a few desirable plants for the pur-

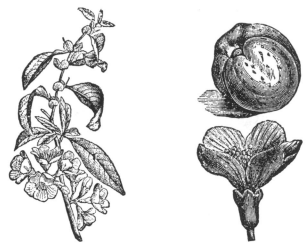

Fig. 113.—BRANCH OF JAPAN QUINCE.   Fig. 114.—FRUIT AND FLOWER.

pose. The Japan Quince, *Cydonia Japonica*, of which figures 113 and 114 show a branch, flower and fruit, is one of the best deciduous plants for an ornamental hedge. It will grow in almost any soil; if left to itself it forms a dense, strong bush, but it may be clipped or trained into any desired form. Its leaves are of dark glossy green, they come early in spring and remain until late in Autumn. This is one of the earliest shrubs to bloom in spring ; its flowers are generally intense scarlet, though there are varieties with white, rose-colored, or salmon-colored flowers. A hedge of this plant is not only highly ornamental, but its abundant thorns make a good barrier. Privet, *Ligustrum vulgare,* makes a very neat screen, but will not bear severe cutting back, and is therefore suitable only for grounds of sufficient extent to admit of its being allowed to make unrestrained growth. The common Barberry, *Berberis vulgaris,* also makes an exceedingly pretty screen in time, but it is of slow growth. The Buffalo Berry, *Sheperdia argentea,* has been tried for hedges, but for some reason it has never attained any popularity. In the Southern States, the Cherokee Rose has been found quite successful for the purpose, and nothing in the shape of a hedge can exceed, in striking effect, one of these in full bloom. For evergreen screens nothing is better than the Hemlock, *Tsuga Canadensis.* The Norway Spruce is of rapid growth and bears cutting well. The Arbor Vitæ, *Thuja occidentalis,* is also very successfully employed for the purpose.

CHAPTER VIII.

PORTABLE FENCES AND HURDLES.

PORTABLE BOARD FENCES.

Figure 118 shows a very strong and secure board fence, composed entirely of ordinary fence boards. The triangular frames, which serve as posts, are each of two pieces of inch boards, crossed and braced as shown in figure 115. The panels, figure 117, are sixteen feet long, each composed of four boards, six inches wide. The space between the lower two boards is two and a half inches, second space three and a half inches. A convenient way of making the panels is to use three horses, like that shown in figure 116, the length of each being equal to the total

Fig. 115.—THE POSTS.

width of the panel, and the three short upright strips marking the respective spaces between the boards. The top is covered with iron to clinch the nails used in putting the panel together. The boards are laid on these horses, and the upright cross-pieces nailed on. The second board from the top of each panel is notched at both ends, as shown in figure 117. A good way to make the trian-

Fig. 116.—"HORSES" FOR MAKING THE FENCE.

gular frames alike, is to cut the pieces by a uniform pattern. Then make one frame of the size and form desired, and at each of the three places where they are nailed together, fasten a plate of iron, thick enough to prevent the penetration of a common wrought nail driven against it. Now lay this pattern frame on the floor with the iron bolts uppermost. Then lay three pieces on this in exactly the right position, drive wrought nails through the two pieces and against the iron plates, which will clinch the nails firmly as fast as they are driven. This will enable the man to nail the frames together quite

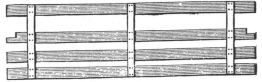

Fig. 117.—A SINGLE PANEL.

rapidly. In setting up the fence, each triangular frame supports the ends of two panels. The upper and lower boards of each panel interlock with the frame, as shown in figure 118, making a very strong fence. On open prairie or other wind-swept situations, it may be necessary to stake down some of the frames, to prevent their blowing over. This is quickly done by sharpening pieces

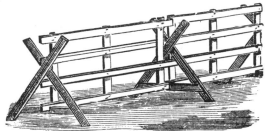

Fig. 118.—THE FENCE IN POSITION.

of inch boards, twelve inches long, and one inch wide, and driving one beside the foot of the triangle, where it rests on the ground, and putting an eight-penny nail through both.

PORTABLE FENCES OF POLES OR WIRE.

Figures 119 and 120 show styles of portable fences, which are used to some extent in the territories. The base of each is the half of a small log, split through the center. For the fence shown in figure 119, two augur holes are bored a few inches apart, and small poles driven to serve as posts. Rails or round poles of the usual length are laid to the desired height, and the top

Fig. 119.—PORTABLE POLE FENCE.

of the posts tied together with wire. In situations where timber is less plentiful, a single stake is set into the base, as in figure 120, braced, and barbed or plain wire attached by staples. Besides the advantage of being

Fig. 120.—PORTABLE WIRE FENCE.

easily moved, these fences can be prepared in winter, when there is little else to do, and rapidly set in place at any time when the ground is clear of snow.

Figure 121 is a fence made of either sawed stuff, or of rails or poles, having their ends flattened and bored. An iron rod, or piece of gas-pipe, any where from half an inch to an inch in diameter, is run through the holes, and through a base block into the ground as far as necessary. A round stick of tough durable wood, an inch or more in diameter, will answer. The size of this rod and its strength will depend upon the amount of zigzag

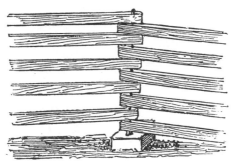

Fig. 121.—PORTABLE FENCE OF POLES OR RAILS.

that is given to the lengths. If the corners are one **foot** on each side of a central line, the fence firmly held together by the rods, will in effect stand on a two feet wide base. Less than this would perhaps sometimes answer, and there are no sharp corners, or deep recesses for weeds and rubbish.

PORTABLE FENCES FOR WINDBREAKS.

A fence that can be easily moved and quickly set up is shown in figure 122. It consists of panels made of strips

Fig. 122.—PORTABLE FENCE.

eight or ten feet long, nailed to two by four posts, which
are beveled to a sharp corner at the lower end. These
panels are supported by posts, placed as shown in the engraving, and pinned to the fence posts by wooden pins,
driven in by a light mallet. The panels are light and
can be loaded upon a wagon from which the sides and
ends of the box are removed. A box of pins and the
mallet are all the tools required to set up the fence. This
fence is not easily overthrown by the wind, which holds
it down firmly when blowing on the face of it. For this
reason in windy localities, the fence should be set facing
the windy quarter.

Another good form of movable fence is seen in figure 123. It is made of common fence-boards, securely

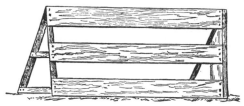

Fig. 123.—RAILROAD WINDBREAK.

nailed on very light posts or on the edge of narrow boards
and braced as shown in the engraving. This style of
panel is largely employed by railroads as windbreaks in
winter to keep the tracks from becoming covered with
drifted snow. It is equally convenient on the farm, when
a temporary inclosure is needed.

PORTABLE POULTRY FENCES.

It is often very convenient when poultry are inclosed
during the growing season, to have a fence for the hen-
yard which can be readily moved from place to place.
The illustration, Figure 124, shows one of these. Cut
the posts the same length as the pickets, and to the inner

side of each attach two strong iron hoops bent into a
semi-circle, one near the bottom and the other half way
up. Through these hoops drive stakes fitted to fill them

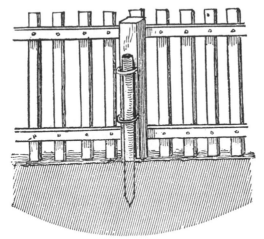

Fig. 124.—PORTABLE POULTRY FENCE.

closely, with sharpened points for easily entering the
ground. When removing the fence the posts can be
slipped off.

Turkeys, even when they have attained a considerable
size, should be shut up until after the dew is off the
grass, and other fowls must be confined in limited runs,

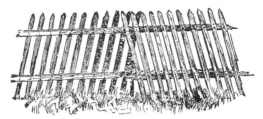

Fig. 125.—MOVABLE FENCE FOR TURKEYS.

while the young are small. It is quite an advantage if
these runs can be changed easily, and this can be accomplished only when they are enclosed in a light movable

fence. Such a fence is shown in figure 125, on preceding page. It is made in twelve or sixteen feet sections by nailing laths to light pieces of the proper length. The upper end of the laths is sharpened ; the end ones are of

Fig. 126.—CROSS-SECTION OF MOVABLE FENCE.

double thickness. The sections are placed with the end-laths intercrossing at the top, and about six inches apart at the bottom, as in cross-section, figure 126. They are held apart by blocks, figure 127, which rest on the upper edges of the cross-pieces and against the laths. They are held together, and to the ground, by stakes driven against the outer side of the end laths. As these stakes have the same angle as the laths, they hold the sections together, and also the fence in its place and down to the ground. The triangular space where the sections join is

Fig. 127.—CROSS-BLOCK FOR FENCE.

closed by a lath driven in the ground or tacked to the block between the cross-pieces. Corners must be formed of two sections inclined inward, and in the same way that sections are joined. The stakes are readily withdrawn, and the sections are so light that they are easily handled.

PORTABLE FOLDING FENCE.

A very convenient form of portable fence or hurdle is illustrated in figures 128, 129 and 130, which was

Fig. 128.—FENCE IN POSITION.

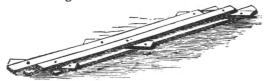

Fig. 129.—FENCE FOLDED.

Fig. 130.—AS A SIDE HILL FENCE.

brought out some five or six years ago. It may be constructed with two or three upright pieces of two-by-four-inch scantling, and four bars, figure 128, held together by carriage bolts in such a manner, that each panel can be closed when desired, as a parallel ruler is folded together. As the bars are on alternate sides, the panel, when closed, takes up the space of two bars only, figure 129. The fence may easily be removed, and fits itself to rolling ground or side-hill, as shown in figure 130. When in position it may be supported by stakes of the same thickness as the upright bars, and driven close beside them.

TEMPORARY WIRE AND IRON FENCES.

Several kinds of wire and iron fences are used in France to make temporary enclosures for exhibition purposes. Two forms are illustrated herewith. Figure 131

Fig. 131.—TEMPORARY WIRE FENCE.

is made of plain iron wire with cast or wrought iron posts. Each post has a plate on its lower end, which is set eighteen inches below the surface of the ground, and the earth filled in compactly about it. The front of the engraving shows the holes in section, with the plates. The top strand is a wire rope made by twisting several strands together. The fence seen at figure 132 is made of narrow strips of sheet iron attached to iron posts driven into the ground. The gate, like that of the other form, is provided with small wheels, which run on a track. The two fences may be modified by using wooden posts sharpened

Fig. 132.—TEMPORARY IRON FENCE.

at the lower end, and driven into the ground, then fastening to them with suitable staples strips of rather broad hoop iron at the top, with plain wire below.

CHAPTER IX.

FENCES FOR STREAMS AND GULLIES.

FLOOD FENCES.

In a situation where a line of fence crosses a stream or a gully liable to be flooded, it is necessary to make special provision for it. A fence extending down near the surface and sufficiently rigid to withstand the current, would arrest the drift wood and other objects floated down on the flood, and soon become a dam. The right kind of a fence must therefore yield to the force of the flood, and renew its position, or be easily re-

placed after it has subsided. Figure 133 is a very effectual flood-gate for a running stream. The posts, *B, B,*
are firmly set on the bank, and a stick of timber, *A,* mor-

Fig. 133.—STRONG FLOOD-GATE.

tised on the top of them. The three uprights, *C, C, C,*
are hinged to the cross-timber, and the boards, *F,* fastened in place by tenpenny steel fence nails. The gate
easily swings with the current, *D.* Figure 134 shows a
form which operates in a similar manner like the other.

Fig. 134.—A CHEAPER FLOOD-GATE.

It consists of two stout posts, five feet high, bearing a
heavy cross-bar, rounded at each end, and fitted into

sockets, in which the bar with gate attached can swing. The construction of the gate is easily seen from the engraving.

The above forms are self-acting, and swing back to their places as the water subsides.   For larger streams, it is necessary to construct fences that give way before the flood, and can be brought into position again when it is over.   One of these, for a stream which is liable to bring down much drift wood, is shown in figure 135.

Fig. 135.—FENCE FOR A FOREST STREAM.

The logs are the trunks of straight trees, about eighteen inches in diameter, which are hewed on two sides ; posts are mortised in each of these logs, and on them planks are firmly nailed.   The logs are then linked together with inch iron rods, and the first one connected by means of a long link to a tree or post firmly set in the ground upon the banks of the stream.   The links must all work freely.   When high water occurs, the fence is washed around and left on the bank ; after the water has subsided sufficiently, the logs may be dragged back to their places, as shown in the engraving, by means of a horse,

hitched to a staple in the end of the log. Figure 136
shows a lighter fence made of poles or rails, held by in-
terlinking staples to the posts on the side of the stream.
As the floods come down, the rails are washed from the

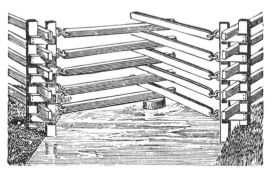

Fig. 136.—FENCE OF MOVABLE RAILS.

center, and float freely at either side of the stream. They
can be laid up in place again when the water subsides.

The fence shown in figure 137, though rather rude and
primitive, has the advantage of being cheaply constructed and permanent. Two strong posts are driven into
the banks on the margin of the stream, to which a log,
a foot or more in diameter, is fastened by pins, spikes or
withes, about twenty inches above low water mark. Then

Fig. 137.—AN EXTEMPORISED FLOOD-FENCE.

fence rails are sharpened at one end, driven into the
stream above the log, upon which the other ends rest,
projecting about a foot. They are then securely spiked
or pinned to the log, and the work is done. The pointed

ends of the rails are up the stream, and in case of flood, the water pours over the obstruction, carrying with it brush, driftwood, etc.

The flood-gate, figure 138, is designed to prevent small stock from passing from one field to another through a water-course under a fence where there is low water, while in time of high water the gate will rise sufficiently

Fig. 138.—AUTOMATIC FLOOD-GATE.

to allow the floating trash to pass through, but not higher, as it is self-fastening.   The invention consists of a gate constructed of perpendicular slats hinged above, and moving. This hangs across a stream or ditch. On the down-stream side of the gate a swing paddle is fixed, which hangs in the water.   This, marked *a* in the illustration, is attached to an angular bar, *b*, which is moved when the flow of water presses with force against the paddle.   Two notched pieces, *c c*, attached to the gate, rest upon the angular bar, *b*, at low water; when both the paddle and the gate hang at rest, perpendicularly, these notched pieces, *c c*, hold the gate firmly shut; when, however, the water rises and the paddle is moved sufficiently to disengage the notches, the gate will be moved by the force of the water, and if sticks or rubbish of any kind float down against it they will be swept under it by the water. When the water subsides, the paddle swings back, the pieces, *c c*, catch and keep the gate closed at any height it may fall to. Let the cross-piece, *d*, that is halved into the posts, be about one foot above the banks of the ditch. The pieces, *f f f f*, represent the fence above the ditch, the small posts, *g g*, with the pieces nailed to them, are to prevent the stock from passing when the gate is partly closed, at the same time bracing the posts, *e e;* the holes at *h* are to raise and lower the paddle *a;* if small, a cleat on one of the arms upon which the piece *B* is hung, prevents the paddle from swinging towards the gate.

Figure 139 shows a kind of fence used in Missouri to put across sloughs. It is in effect two panels of

Fig. 139.—A MISSOURI FLOOD-FENCE.

portable fence. The posts are set three to four feet deep, with the tops about one foot above ground ; the other posts, to which the planks are nailed, are bolted to the top of the inserted posts. The ends of the panel that connect with the post on the bank are slightly nailed with cross-strips near the top, so as to be easily broken loose when the flood comes. There are also temporary braces bearing upstream, put in to prevent the fence from falling, but are easily washed out, when the fence falls down stream, and logs and other obstructions pass by readily. As soon as the flood goes down, the fence is easily raised, a panel at a time, to a proper place.

Figure 140 shows a cheap and effective form of flood fence. The material used are square-hewn timbers, seven

Fig. 140.—FRESHET FENCE.

or eight inches for sills, stone pillars, split rails about ten feet long. The rails are driven in the ground about two feet deep; the upper ends project above the sill two or three feet, and are spiked down to the sill with large spikes; when the freshet comes, logs and drift-wood are carried over, and the fence will be left in as good order as before the high water.

Figure 141 represents a gulch fence or gate which is in common use in some parts of the Pacific Slope. It

Fig. 141.—CALIFORNIA GULCH FENCE.

is particularly adapted to the gulches of the foot hills and the irrigating ditches of the plains. The whole gate swings freely by the upper pole, the ends of which rest in large holes in posts on either bank, or in the cross of stakes. The upright pieces may be of split pickets or sawed lumber, as may be the most convenient. If the stream is likely to carry floating brush, logs, etc., the slats should be of heavier material than is necessary when this is not the case. When constructed properly the gate will give, allowing rubbish and freshets to pass, and then resume its proper position. The principal advantage claimed for this gate is that it is not apt to gather the passing debris.

A gully is sometimes difficult to fence properly, but by hanging a frame over it, as is seen in figure 142,

Fig. 142.—FENCE FOR A DRY GULLY.

the object may be quickly accomplished. The frame can be spiked together in a short time, or framed together if a more elaborate one is desired. To make it serve its purpose completely, the rails must be closer together near the bottom than at the top of the frame, in order to prevent small animals from going through it.

A modification of this last named device, seen at figure 143, gives greater space for the passage of brush wood or other large objects, which may be swept down on the flood. The width, strength and size of the bases supporting the side posts, and of the braces, will depend upon the width and depth of the channel. The base pieces can be firmly anchored by stakes driven slanting over the ends and outsides, or by stones piled on. For wide, shallow streams, three or even more braced uprights can be anchored eight or ten feet apart in the bed with heavy stones, with two or more swinging sections.

Fig. 143.—A FRESHET FENCE.

If small trees or long timbers are likely to float down, the swinging gate may be twelve or fifteen feet wide. For smaller streams, with strong high banks, five or six feet will suffice.

PORTABLE TIDE FENCE.

Figure 144 represents a fence for tide-creeks. It is made usually of pine, the larger pieces, those which lie

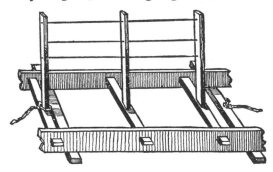

Fig. 144.—SECTION OF A TIDE FENCE.

on the ground and parallel with the run of the fence, are three by four-inch pieces, hemlock or pine, and connected by three cross-bars, of three by four-inch pieces, mortised in three feet apart. Into the middle of these three cross-pieces, the upright or posts are securely mortised, while two common boards are nailed underneath the long pieces, to afford a better rest for the structure, when floating on the water or resting on the ground. Barbed or plain wires are stretched along the posts, which are four feet high.

WATERING PLACE IN A CREEK.

Cattle naturally select a certain place in a water-course to drink at, where the bank is not precipitous. During a good part of the year this bank is muddy, on account

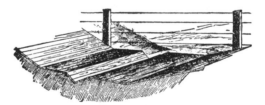

Fig. 145.—A CLEAN WATERING PLACE.

of its moisture and trampling of the animals. As a result, the horses get the scratches, the cows come to the milking pen with muddy udders, and frequently animals are injured by the crowding in the mud. Hogs are often seriously injured, because the mud becomes so deep and tough, that they are well nigh helpless in it. Another objection is that the animals wade to the middle of the creek, and soon make its bottom as muddy as the bank, and the water becomes unfit for drinking. The arrangement shown in our illustration, which may be built of heavy plank, brick, or flat stones, prevents all this. It

is constructed by first making an incline to a level plat-
form for the animals to stand on while drinking. This
plane terminates in an abrupt descent, forming a trough
for the water to flow through. The trough should not
be more than two feet wide, that the animals may easily
get across it. The level floor permits the animals to
drink at their ease, often a matter of importance. Such
a drinking place should be made at the upper end of the
creek, where it passes through a field to prevent the animals from soiling the water by standing in it above where
they drink.

CHAPTER X.

MAKING AND SETTING POSTS.

MAKING FENCE POSTS.

There is quite an art in splitting logs into posts.
Every post should have some heart wood, which lasts the
longer, for two reasons : That there may be durable
wood into which to drive the nails, and without it some
of the posts, composed entirely of sap-wood, will rot
off long before others, making the most annoying of
all repairing necessary. If the log is of a size to make
twelve posts, split along the lines of figure 146, which will
give each post a share of heart wood. This will make a
cross section of the posts triangular, the curved base being somewhat more than half of either side. This is a
fairly well shaped post, and much better than a square one
having little or no heart wood. Although the log may
be large enough to make sixteen or eighteen posts, it is
better to split it the same way. It should first be cut
into halves, then quarters, then twelfths. If it is at-

tempted to split one post off the side of a half, the wood will "draw out," making the post larger at one end than the other—not a good shape, for there will be little heart wood at the small end. When the log is too large to admit of it being split in that way, each post may nevertheless be given enough heart wood by splitting along the

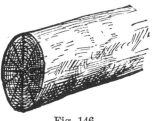
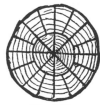

Fig. 146.　　　　　　　　　　　　Fig. 147.

lines, shown in figure 147. First cut the logs into halves, then quarters, then eighths. Then split off the edge of each eighth, enough for a post—about one-fourth only of the wood, as it is all heart wood, and then halve the balance. A good post can be taken off the edge, and yet enough heart wood for the remaining two posts remain.

A POST HOLDER.

A simple arrangement for holding a post while it is being bored or mortised, is shown in figure 148. It con-

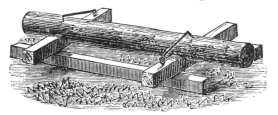

Fig. 148.—A POST HOLDER.

sists of two long pieces of round or square timber, lying parallel upon the ground, and two shorter sticks resting

upon them at right angles. The upper pieces have sad-
dles cut out for the posts to fit into. A staple with a
large iron hook or "dog," is fastened into one end of
each cross-piece, as shown in the engraving. When the
post is laid in position, the hooks are driven into it
holding it firmly.

———◆◇◆———

DRIVING FENCE POSTS BY HAND.

Where the soil is soft, loose, and free from stone, posts
may be driven more easily and firmly than if set in holes

Fig. 149.—DRIVING FENCE POSTS.

dug for the purpose. An easy method of driving is shown
in figure 149. A wagon is loaded with posts and furnished with a stage in the rear end of the box, upon
which a person can stand to give the posts the first start.
Another man holds the posts upright while they are

driven. When one post is driven to its place, the wagon is moved to the next place, and this operation repeated.

To drive posts, a wooden maul should be used. This is made of a section of an elm trunk or branch, eight or nine inches in diameter, figure 150. An iron ring is driven on each end, and wedged all around, the wood at the edge being beaten down over the rings with a hammer or the poll of an axe. To prevent the posts from splitting or being battered too much, the ends of the maul should be hollowed a little, and never rounded out, and

Fig. 150.—MAUL FOR DRIVING POSTS.

the ends of the posts should be beveled all around. The hole in the maul for the handle should be made larger on one side, and lengthwise of the maul, and the handle spread by two wedges driven in such a way as not to split the maul.

TO DRIVE POSTS WITHOUT SPLITTING.

Posts are very liable to split in driving, unless some precaution is used. This damage and loss can be avoided in a great measure by proper preparation of the posts before they are driven. The tops of sawed posts should have the sides cut off, as in figure 151, or simply cut off each corner, as in figure 153, while a round post should be shaped as in figure 152. The part of the post removed need not be more than half an inch in thickness, but when the corners only are cut away, the chip should be thicker. In driving, it is very important to strike the post squarely on the top, and not at one corner or

side. In most soils at the North, the frosts heave posts
more or less each season, and they need to be driven
down to the usual depth. To do this with little injury to the post, the device shown in figure 154 may

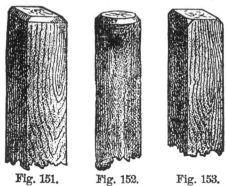

Fig. 151.   Fig. 152.   Fig. 153.

be used. It is a piece of tough hard wood scantling, *e*,
eighteen inches in length, with tapering ends. It is
provided with a handle, *h*, three feet in length, of quite
small size, and if possible, of green timber. In using it,
let one person (a boy will do) lay the bit of scantling on
top of the post to be re-driven, when, with the beetle or

Fig. 154.—SCANTLING WITH HANDLE IN POSITION.

sledge, the scantling, instead of the post is struck, thus
preventing the splitting of the post. When the top of a
fence is surmounted by a stringer, as in the engraving,
the effect of the blow is distributed over a large space,

and both stringer and post escape injury. The attendant should keep hold of the handle, *h,* while the posts are being driven, and move the scantling from post to post as required.

A POWERFUL POST DRIVER.

For a farmer who has a large number of posts to set, a special apparatus for driving them will be found useful.

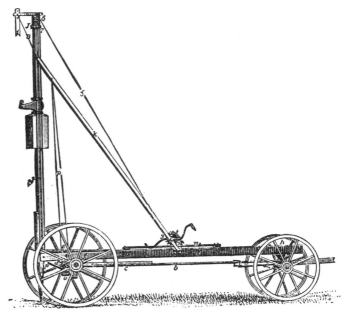

Fig. 155.—THE POST-DRIVER.

The accompanying illustrations show a machine of this kind. An axle, *a,* figure 155, of hard wood, eight and one-half feet long; a hickory sapling will do. It has spindles shaved to fit the hind wheels of a wagon, which are fastened by linch-pins, leaving about six feet space between the hubs. A coupling-pole, *b,* thirteen feet long, is framed in and strongly braced at right angles with the axle, and connects in front with the forward axle of a common wagon. The main sill, *d*, figure 156, is one stick of timber, six by eight inches, by fourteen feet long and has a cross-piece, *e*, framed in the end. Two side-pieces, *f*, two by four inches by five feet long, are pinned or bolted to the main sill at *g*, and cross-pieces framed into them, as shown in figure 156, so framed that the lower edges of the side-pieces will be two inches from the axle, when the main sill rests on the axle. The side-pieces, *f*, should be twenty-two inches apart at the ends. The front end of the main sill rests on the front axle, in place of a bolster, and the "king-bolt" passes through it at *h*; the upright guides, *i*, are two by four inches by fourteen feet long, bolted to the side-pieces, *f*, with a space of fourteen inches between; a cap, *j*, two by three by twenty-six inches long, is framed on top. Two braces, *k*, two by four inches by sixteen feet long, are bolted to the upright guides, two feet below the cap, and connect at the bottom with a cross-piece, *l*, two by eight by twenty-two inches long, between the braces. It has rounded ends passing through two-inch holes in the braces, and fastened by a pin outside, to form a loose joint. This cross-piece, *l*, is held down on the main sill by a strip, *m*, and steadied by cleats; it is free to slide back or forward, and is held in place by a short pin. By moving this cross-piece, the upright guides, *i*, are kept perpendicular when going up or down hill. A small windlass, *o*, figure 155, is placed under the axle, *a*, between hangers framed into the axle, close to the hubs. Two brace-ropes, or wires, *p*, are fastened to this windlass at the extreme ends, and wound around it a turn or two in opposite directions, drawn tight and fastened to the main braces near the top. By turning the windlass, *o*, slightly, by means of a short bar, the machine may lean to either side, to conform to sliding ground, thus being adjustable in all di-

rections. The maul, *r*, figure 157, of tough oak, fourteen
by eighteen inches, by two feet long, weighs about two
hundred pounds, is grooved to fit smoothly between the
guides ; the follower, *s*, is more plainly shown in the engraving, also the simple latch, by which the follower and
maul are connected and disconnected. The square clevis, *t*, is of three-quarter inch iron, suspended from the

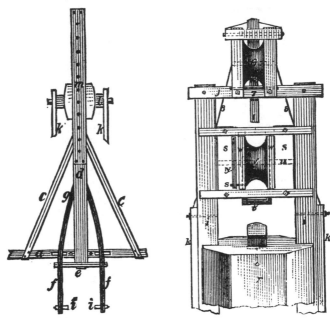

Fig. 156.—BOTTOM OF DRIVER.   Fig. 157.—TOP OF UPRIGHT.

same iron pin, *u*, on which the pulley, *v*, is placed. It
is partly imbedded in the wooden casing, *w*, which is
eight by eighteen inches ; this casing serves to inclose
the pulley, *v*, and also to trip the latch when brought together ; the clevis, *t*, is caught under the hook fastened
in the maul, is pressed into place by a small hickory
spring, *y*, acting on a small iron pin, *z*; when it reaches
the top, the crotch, 1, suspended from the top, comes
in contact with the pin, 2, and the clevis, *t*, is pressed

back, and releases the hook, *x*, when the maul drops.
The windlass, 3, figure 155, has two cranks, and a
ratchet for convenience.   The rope passes from the windlass over the pulley at the top, down and under the pulley, *v*, then up, and is fastened at 7, on the cap, *j*,
wire braces at 8.   By releasing the cranks and ratchet,
the follower will run down the guides, and, striking the
maul, will "click" the latch into place, ready for another hoist.   For two men it is easy work, and can be
handled quite rapidly.   Drive astride the proposed line
of fence ; lay a measuring-pole on the ground to mark
the spot for the next post ; drive forward with the post-
driver, having the maul partly raised, set up a post, and
proceed to drive it.

SETTING A GATE POST.

No matter how strong or how well braced a gate may
be, it will soon begin to sag and catch on the ground, if

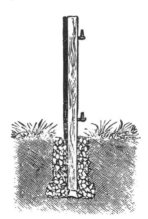

Fig. 158.—A GATE POST SET IN CEMENT.

the gate post is not firmly planted.   Sometimes, owing
to the soft nature of the ground, it is almost impossible

to plant the post firmly, but in such cases the work can
generally be done satisfactorily by packing medium-sized
stones around the post, in the hole, as shown in figure
158.   If it is thought that this will not insure sufficient firmness, add good cement.   Place in a layer of
stones, then cement enough to imbed the next layer of
stones, and so on, until the hole is full and the post
planted.   Do not cover up the stones with earth or dis-
turb the post for a few days, until the cement has "set."
Remember that the post must be set plumb while the
work is going on, as it can never be straightened after
the cement has "set."   Only durable posts should be
used, and this method of setting should only be followed
with gate posts which are supposed to be permanent, and
not with posts liable to be changed.

A still better method is shown in figure 159.   Before
the post is set into the hole, a flat stone is laid edgewise

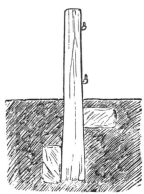

Fig. 159.—GATE POST BRACED WITH STONES.

in the bottom, on the side which is to receive the great-
est pressure from the foot of the post.   When the post is
set, and the hole half filled with earth, a second stone is
placed against the post on the side to which it will be
drawn by the weight of the gate.   The stones receive
the pressure and hold the post firmly in position.

FENCE POSTS FOR WET LANDS.

Low meadow and other marsh land is subject to heaving by the frost, and much difficulty is experienced in securing firm fences upon such ground, as the posts are drawn up by the freezing of the surface. To avoid this, much may be done in the way of selecting posts that

Fig. 160.—DIFFERENT METHODS OF TREATING POSTS.

are larger at one end than the other. It will help very much to put a strong, durable pin through the bottom end of the post, or to notch it at each side, as in figure 160, and to brace the bottom with a flat stone, driven well into the side of the hole with the rammer. When the soil is very soft and mucky, it is best to drive the posts and to make them hold well in the ground, to spike wedge-shaped pieces to them on either side, by which they are held firmly in their places.

LIVE POSTS.

A living tree which stands in the right place, makes a very durable and substantial fence-post. In the great treeless regions of the Mississippi Valley, where it is difficult to obtain timber for posts, it is not an unusual practice to plant trees for the purpose on street boundaries, and other places where the fence is to be permanent. White willow is well adapted for the purpose on suitable soils, as it grows rapidly and bears close pruning. In situations where the soil is even moderately damp, white willow posts, four inches in diameter, cut green and set

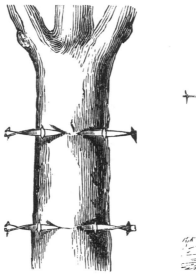
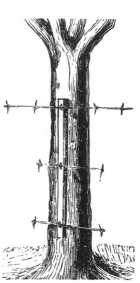

Fig. 161.   Fig. 162.

in spring, will take root and grow.   The new branches soon form a bushy head, which may be cut back from time to time.   It is not advisable to nail boards or drive staples directly into the tree.   With a board fence, the swaying of the tree loosens the nails, and if barbed wire is stapled to the tree, the bark and wood will in time grow over them as in figure 161.   To obviate this, a stick is nailed to the tree as in figure 162, and to this the fence is attached.   A still better method is to secure the strip of wood to the tree by two or three pairs of inter-locking staples.

———◆◇◆———

MENDING A SPLIT POST.

Fence posts split from a variety of causes, and when they are in this condition they make a very insecure

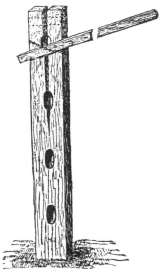

Fig. 163.—MENDING A SPLIT POST.

fence. The usual way is to merely nail an old horseshoe or two across the split part, just below the holes in the posts. This answers fairly well, but does not draw the cleft together, and horseshoes are not always on hand. A better method of doing this is shown in figure 163. A short, stout chain is put around the top of the post, just tight enough to admit of a strong lever. The parts of the posts are then brought together by a heavy downward pressure of the lever and held there, while a strip of good tin, such as can be cut from the bodies of tin cans, is put around and securely nailed. If the post is a heavy one and the cleft large, it is well to take the entire body of a can and double it, to give it additional strength before nailing it on. The dotted lines show where the tin is nailed.

HOOK FOR WIRING POSTS.

Figure 164 shows a modified cant-hook for drawing together the upper extremeties of fence stakes that are to be

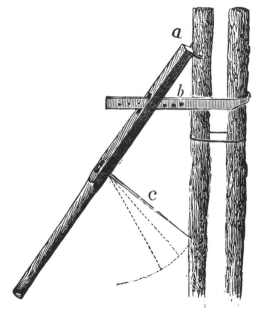

Fig. 164.—A STAKE DRAWER USED IN WIRING FENCES.

wired, as in the engraving. The half-moon shaped iron, *a,* is riveted fast to the top end of the lever, and is to prevent the end of the lever from slipping off the stake when in use. The second iron from the top, *b,* is twenty-five inches long, with two hooks at the end, though one will do ; this is to catch the stake on the opposite side of the fence. This iron is fastened in the lever by a bolt in a long mortise, in the same way, as the hook in an ordinary cant-hook. The iron rod, *c*, has a hole in one end, and is drawn out to a point at the other—this is fastened to the lever by a bolt in a long mortise, and serves to catch in the stake or rail, and hold the stakes together, while the man adjusts the iron around the stakes. When the stakes are drawn tightly to the fence, this rod is drawn up until it strikes the stake or one of the rails, when the man can let go of the "drawer," and it holds itself. The lever is four feet and three inches long, and two inches square, with the corners taken off part of the way down, the lower end being rounded for a handle, as shown in the engraving.

DRAWING FENCE POSTS.

Figure 165 shows a practicable method of drawing out fence posts by the aid of an ox team. A stout piece

Fig. 165.—DRAWING FENCE POSTS.

of timber with a large flat "foot" is placed under the chain to change the direction of the draft. Two men and a steady yoke of oxen can extract fence posts very

quickly and easily by this method. A good steady team
of horses will do quite as well as oxen.

LIFTING POSTS BY HAND.

A convenient and sensible implement, for taking up
fence posts without the aid of a team, is shown at figure
166. It consists of a stout pole of the size and shape of

Fig. 166.—A CONVENIENT POST LIFTER.

a wagon tongue. The thicker part of this pole, for about
fifteen inches from the end, is shaped into a wedge.
This is sheathed with a frame made of iron, half an inch
thick and two and a half inches wide, and securely fastened with screws or bolts. The end should be pointed
and slightly bent upwards. The manner of using this
convenient implement is shown in the illustration.

Frequently a farmer has occasion to lift posts, and has
not time to wait for the construction of an iron-shod lever. Figure 167 shows a very simple, inexpensive contrivance for such cases. A spadeful of earth is taken
from each side of the post, and a short, strong chain
loosely fastened around the lower end of the post, as far

down as it can be placed. A strong lever—a stout rail will answer the purpose—is passed through the chain, as shown in the engraving, until the end of the rail catches firm soil. By lifting at the other end of the lever the post is raised several inches, when both chain and lever are pushed down again for a second hold, which general-

Fig. 167.—LIFTING A POST.

ly brings the post out. The chain is furnished with a stout hook at one end, made to fit the links, so that it can be quickly adjusted to any ordinary post.

SPLICING FENCE POSTS.

There are places, as crossing over gullies, etc., where unusually long posts are desirable, though not always easy to obtain. In such cases properly spliced posts are almost as durable as entire ones. The engraving of the front and side views, figure 168, shows how the splice may be made to secure strength and durability. The splices should be made with a shoulder at the lower end, and well nailed together, after which one or two bands of hoop-iron may be passed around the splice and

Fig. 168.—SPLICING FENCE POSTS.

securely fastened. The hoop-iron band is one of the most important points in a splice of this kind.

APPLICATION OF WOOD PRESERVATIVES.

To prevent decay at the center, as well as of all that part of the post placed below ground, by use of wood preserving solutions, the following system is both novel and valuable : It is to have a hole in the center of the post, from the bottom upward, to a point that shall be above the ground when the post is in position. Then bore another hole in the side of the post with a slight inclination downward, making an opening in the center hole, as shown in figure 169. A wooden plug, two or three inches long, should be driven snugly into the hole at the bottom of the post, in order to prevent the escape of any liquid that may be used in the operation. When the posts are set in an upright position, a preservative solution may be introduced into the hole in the side and the centre one filled with it, after which a cork plug of some kind should be inserted in the side hole, to prevent evaporation, as well as to keep out dust and insects. The solutions thus introduced will gradually be absorbed by the surrounding wood, until all parts along the entire length of the central cavity must become completely saturated. When the solutions used have been taken up by the surrounding wood, it will only be nec-

Fig. 169.—SECTIONAL VIEW OF BORED POST.

essary to withdraw the cork or plug, and apply more, if it is thought desirable. A common watering pot with a slender spout will be a handy vessel to use in distributing the solutions.

Petroleum, creosote, corrosive sublimate, or any other of the well known wood preservatives may be used in this way. Telegraph posts might be prepared in the same way, and if the central reservoirs were kept filled with petroleum, they would last a hundred years or more. Where a large number of posts or poles are to be prepared, it would be cheaper to have the holes bored by steam or horse power than by hand. With very open and porous wood it is quite probable that a hole bored in the side of the post and above the ground, and deep enough to hold a half pint or more of creosote or some other similar solution, would answer, but a central cavity reaching to the bottom, would perhaps, be best.

IRON FENCE POSTS.

The advent of wire fences was followed by a call for posts in the prairie regions, where timber is scarce. Sev-

Fig. 170.—POST.

Fig. 171.—DISC.

eral forms of iron posts have been devised, of which the leading ones are illustrated herewith. Figure 170 is of iron, one quarter of an inch thick and two and a half inches wide, rolled to a curve and pierced at the proper intervals for the staples, which are to be clinched on the concave side. The disc, figure 171, is swedged out of one fourth inch iron. It is sunken a little below the ground, and the post driven through the curved opening, into

which it fits closely. Figure 172 is a flat iron bar, with
slots cut diagonally into one side to receive the wire.
The post is supported by two tiles with holes to fit the
post, which is thrust through them.

Figure 173 is made of angle iron braced at the surface
of the ground, with an angular iron plate rolled for the
purpose, and driven to its place. Figure 174 shows an

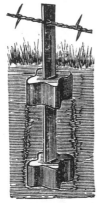
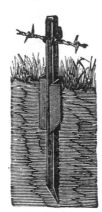

Fig. 172.—POST WITH TILES.   Fig. 173.

iron post, with the ground-piece and driving tube to the
left of it. The post is a round iron bar or tube, with
notches for the wires, which are held in place with short
pieces of binding-wire, wound around the post. The
ground piece, which is shown in the middle of the engraving, is of cast iron, eleven inches long, and five inches
across the top, with two loops for inserting the iron post.
This is driven into the ground, and the iron post driven
through it. At the left of the engraving is shown the
device for driving the post. It is a piece of common
gas-pipe, just large enough to slip easily over the top of
the post, and provided on the top with an iron cap to
receive the blow of the large hammer or maul used in
driving. Figure 175 shows a cast iron ground piece, and
at the right is the lower end of a post resting in one of

them.   The three flanges are cast in one solid piece,
with a hole through the centre of any desired form and

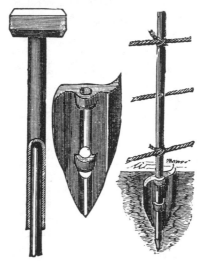

Fig. 174.—POST WITH IRON GROUND PIECE.

size.   The wings or flanges are three inch plates, running
to sharp edges on the bottom, so that they can easily be
driven into the ground.   They may be of any desired
size, larger sizes being required for a light yielding soil

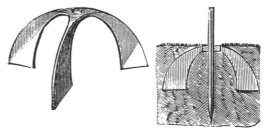

Fig. 175.—CAST-IRON GROUND-PIECE.

than for a stiff one.   Figure 176 is an iron post on a
wooden base, for situations where the ground is soft and
wet.   The base is preferably of cedar, three to four feet
long, four inches thick, and four to six inches wide.   It

is to be sunken in the ground cross-wise with the line of fence. The post is of iron, set and stapled into the end-piece, as shown in the engraving. Before being put in place, the whole is saturated with hot coal tar, as a preservative. There is less call for iron posts than was an-

Fig. 176.

ticipated when wire fences first came into general use. It is found that wooden posts can be delivered in any location reached by railway at less cost than iron posts.

CHAPTER XI.

GATES AND FASTENINGS.

WOODEN GATES.

As board and picket fences have gradually replaced rail and other primitive fences, useful but inconvenient "bars" have begun to disappear, and tidy gates are seen. The saving in time required to take down and put up bars, rather than open and close gates, amounts to a good deal. A good wooden gate will last a long time. Gate-ways should be at least fourteen feet wide. All the wood used in the construction of the gate should be well seasoned. It is best to plane all the wood-work, though this is not absolutely necessary. Cover each tenon with

thick paint before it is placed in its mortise.   Fasten the
brace to the cross-piece with small bolts or wrought nails
well clinched.   Mortise the ends of the boards into the
end posts, and secure them in place with wooden pins
wedged at both ends, or iron bolts.   The best are made
of pine fence-boards six inches wide ; the ends should be

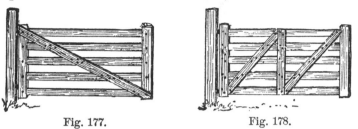

Fig. 177.   Fig. 178.

four by twenty-four inch scantling, although the one at
the latch may be lighter.   Five cross-pieces are enough.
The lighter the gate in proportion to strength, the better
it is.   There is but one right way to brace a gate, and
many wrong ones.   The object of bracing is to strengthen
the gate, and also to prevent its sagging.   Gates sag in
two ways ; by the moving to the one side of the posts

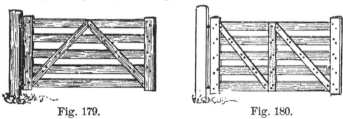

Fig. 179.   Fig. 180.

upon which the gates are hung, and the settling of the
gates themselves.   Unless braced the only thing to hold
the gate square is the perfect rigidity of the tenons in the
mortises ; but the weight of the gate will loosen these,
and allow the end of the gate opposite the hinges to sag.
It is plain that a brace placed like that shown in figure 177
will not prevent this settling down.   The only opposition
it can give is the resistance of the nails, and these will

draw loose in the holes as readily as the tenons in the
mortises. A brace set as shown at figure 178 is not much
better, as the resistance must depend upon the rigidity
of the upright piece in the middle, and the bolts or nails
holding it will give way enough to allow the gate to sag.

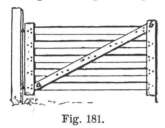

Fig. 181.

Fig. 182.

The method shown in figure 179 is fully as faulty, while
the form shown in figure 180 is even worse. It seems
strange that any one should brace a gate in these ways,
but it is quite frequently seen attempted. The only
right way to brace a gate is shown in figure 181. The gate
may be further strengthened as shown in figure 182. Before the gate can sag, the brace must be shortened ; for

Fig. 183.

Fig. 184.

as the gate settles, the points *a* and *b* must come closer
together, and this the brace effectually prevents.

The posts should be set in such a way that they will
not be pulled to one side and allow the gate to sag. The
post should be put below the line of frost, or else it
will be heaved out of position ; three feet in the
ground is none too deep. Have a large post and

make a big hole for it. Be careful to set the post plumb and stamp the earth firmly in the hole—it cannot be stamped too hard. While stamping, keep walking around the post, so that the earth will be firmed on all sides. Blocks may be arranged as shown in figure 183 ;

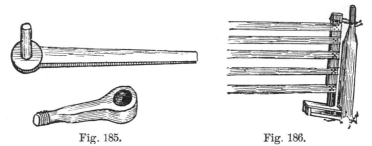

Fig. 185.   Fig. 186.

but this is not really necessary, when the posts have been rightly set, although it may be advisable to take this further precaution.

To remove the pulling weight of the gate when closed, the swinging end may rest upon a block ; or a pin in-

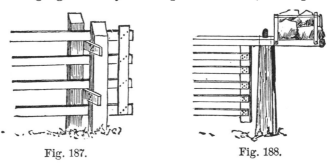

Fig. 187.   Fig. 188.

serted in the end piece of the gate may rest in a slot sawed in the post, or on a shoulder of the post. Figure 184 shows one end of a combination of two plans—the iron rod from near the top of the high post holds the gate while the strain upon the post is lessened by the opposite end of the closed gate being supported on the other post.

For hanging the gate the best hinges are doubtless those shown in figure 185. One part passes through the end-piece of the gate, and is secured by a nut on the end. The other piece is heated and driven into the post, following the path of a small augur-hole. Next to this comes the strap hinge, which should be fastened with bolts or screws. Three easy, cheap ways of supporting the gate are shown in figures 186, 187, and 188. In figure 186, a stout band of wood, or one of iron, may be used in place of the chain. And in place of the stool for the reception of the lower end of the end-piece, a block resting on the ground, or a shoulder on the post, may be substituted. The mode shown in figure 187 is common in the West. Its construction needs no explanation. By sliding the gate back until it almost balances it may be carried around with ease. In figure 188, the fastening, or latch, must be so arranged as to hold the lower part of the gate in position. The box of stone renders it easier to move the gate. A heavy block of wood serves the same purpose.

A VERY SUBSTANTIAL FARM GATE.

Figure 189 shows a gate which combines great durability with much rustic beauty. The cedar posts, *A A*, should be four feet in the ground, and at least ten feet out of the ground. *B* represents a piece of 2 by 6 hard pine, into which the posts are mortised. *C* is a 4 by 4 clear pine, turned at both ends and mortised as shown in figure 191. *D E F* are 1 by 4 pine strips. *G* is a 1 by 6 pine strip, a sectional view being given in figure 190. It is best to use one piece each of *D* and *E*, letting *F* come between them, as it gives more stiffness to the gate. *H* is a block of cedar with a hole bored or dug large enough to receive the post, *C*, and to make it more lasting, a small hole should be bored through the block, so as to let whatever water collects in it pass away; the block should not be less than eighteen inches long—four inches above ground. *I* shows wire fence connected. *J* is a strong wire carried

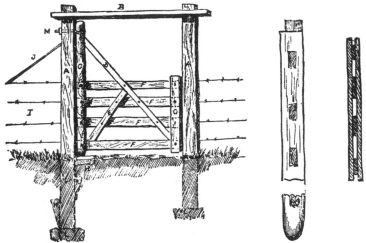

Fig. 189.—A SUBSTANTIAL GATE.   Fig. 190. Fig. 191.

and secured to the bottom of the first fence post. *K K* are cleats attached to posts to keep them more firmly in the ground. *L* are stones for posts, *A A*, to stand on. *M* shows the hinge, made so as to take up the sag after the gate settles, and as the wood wears out.

A STRONG AND NEAT GATE.

The posts, *a, a*, figure 192, of oak or other durable wood, are eight inches square, and stand five and one half feet above the ground. The posts, *b, b*, three and one third inches thick, four and three quarter feet long, are mortised to receive the slats, *c, c*, which are of inch stuff, three inches wide and ten feet four and three-quarter inches long. They are let into posts, *b, b*, at the distance marked in the engraving. The slats, *d*, are three inches wide, and one inch thick, and are placed opposite each other on front and back of the gate as braces ; *e, e,* are simply battens to make a straight surface for the hinges, *f, f;* all except the upper and lower ones are very short and carried back to the post. The hinges, made by a blacksmith from an old wagon tire, are one and one-half inch wide, three-sixteenth inch thick, and are fastened by light iron bolts through the battens at *e,* and to the rear post.

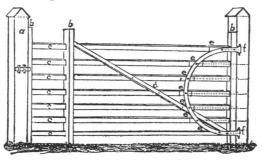

Fig. 192.—A WELL-MADE GATE.

The above describes a cheap, light, durable gate, which in over twenty-three years' use has never sagged, though standing in the thoroughfare of three farms, and also, for years past, used for access to a sawmill. It is made of the best pine. The hinge is an important point. It is not only cheap and easily made, but acts as a brace for the gate at every point, and thus permits the gate to be lightly made. With this hinge sagging is impossible. A gate of this kind will rot down first.

LIGHT IRON GATES.

The gate shown in figure 193 may be made of wrought iron an inch and a half wide and half an inch thick, or preferably of iron gas-pipe of any diameter from half an inch to an inch.   In the vicinity of the oil-regions, pipe can be bought very cheaply, which is in condition good

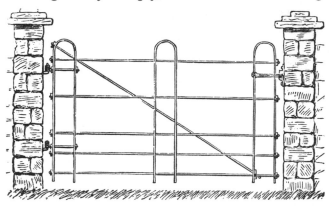

Fig. 193.—A LIGHT IRON GATE.

enough for this purpose.   For guarding against hogs, it should be hung near the ground, and have one or two more horizontal pipes near the bottom.

Figure 194 shows the construction of a gate intended for situations much exposed to trespassers.   It is made

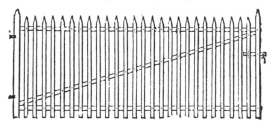

Fig. 194.—A WROUGHT IRON GATE.

of upright strips of flat iron, pointed at the top, and fastened by rivets to a stout frame-work of iron.   The "pickets" are placed two to three inches apart, as desired, for the appearance of the gate, or according to the size of the poultry or animals to be kept from passing.

SELF-CLOSING GATES.

Every self-closing gate should be provided with a drop or spring catch, a suitable bevel for it to strike against and notch to hold it. Gates opening into the garden or out upon the street, should be so hung that they will swing either way. Figure 195 shows a hinge and slide for such a gate. In opening the gate from either side, the arm of the upper hinge slides upon the iron bar, raising the gate a little as it swings around. When loosed, it

Fig. 195.—HINGE AND SLIDE FOR GATE.   Fig. 196.

slides down without help, and closes by its own weight. Figure 196 shows another form of the iron slide, suitable for a wide gate post, and more ornamental than the plain slide in figure 195.

Figure 197 shows a very good and common hanging. The upper hinge consists of a hook in the post and a corresponding eye in the hinge-stile of the gate. The lower hinge is made of two semi-circular pieces of iron, each with a shank, one of which is shown above the gate in the engraving. They are made to play one into the other. This style of hanging may be used on any ordi-

Fig. 197.

nary kind of gate, but is specially useful for a small street gate opening into a door-yard.

There is a style of gate for foot-paths, which is not uncommon, that keeps itself always closed and latched, by means of a single upper and double lower hinge, which

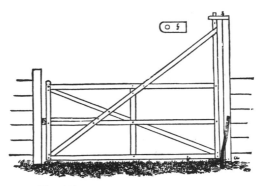

Fig. 198.—SELF-CLOSING FARM GATE.

are to be obtained at most hardware stores. The lower hinge has two "thumbs," which are embraced by two open sockets. When the gate is opened, it swings upon one socket and its thumb, and being thrown off the center, the weight of the gate draws it back, and swinging too, it latches. A farm gate, entirely home-made, may be constructed, of which figures 198 and 199 show the gate and the hinge. The gate is braced and supported by a stay-strip, extending to the top of the upright, which forms the upper hinge, *f* being attached to the top of the gate-post, by an oak board with a smooth hole in it. The lower hinge is separately shown at figure 199. It consists of an oak board, *c*, an inch and a half thick, into

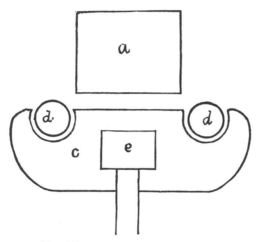

Fig. 199.—LOWER HINGE OF GATE.

which the upright, *e*, is mortised. In this, two sockets are cut, a foot from center to center. The sockets in this case are three inches in diameter, and when the gate is in place and shut, they fit against two stakes of hardwood (locust), two and a half inches in diameter, *d*, which being curved, are nailed to the gate-post, *a*. A smooth stone, laid across in front of these stakes, takes the weight of the gate, and relieves in a measure the pressure on the top of the post. The hinges must be kept well greased, and it is well to black-lead them also, to prevent creaking.

GATE FOR VILLAGE LOT.

Figure 200 shows a light, strong gate made of wood and wire. The top wire is barbed to prevent stock from pressing against it, and beaux and belles from hanging

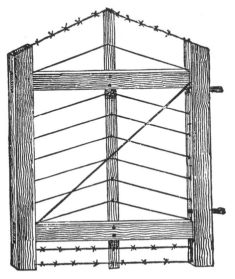

Fig. 200.—CONVENIENT GATE.

over it. The bottom wires are also barbed to prevent cats, dogs, and fowls from creeping under. This gate is cheap, may be easily constructed, and is suitable for either front or back yard.

— ✦ —

A CHINESE DOOR OR GATE SPRING.

Figure 201 shows the manner in which the Chinese use a bow as a spring for closing the light doors and gates. The bow is fastened to the gate by a cord or chain. Another cord or chain is attached to the middle of the bow-string by one end, and the other end is made fast to the gate post, in such a manner that when the

gate is opened, the bow will be drawn, and its elasticity
will serve to shut the gate when released. Our artist has

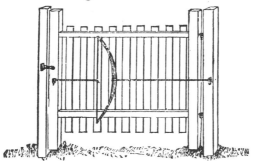

Fig. 201.—CHINESE DOOR OR GATE SPRING.

shown the Chinese invention attached to a gate of Yan-
kee pattern.

LIFTING GATES.

There are various forms of gates not hung on hinges
at all, but either suspended from above to lift, and pro-

Fig. 202.—GATE SHUT.

vided with counterweights, or made in the form of movable panels. Figure 202 represents a gate for general use,

which is peculiarly well adapted to a region visited by
deep snows in winter. The post, firmly set, extends a
little higher than the length of the gate. In front of
this and firmly fastened to it at bottom and top, is a
board at sufficient distance from the post for the gate to
move easily between them. An iron bolt through the
large post and the lower end of the tall, upright gate
bar, serves as a balance for the gate to turn on. A
rope attached to the bottom of the gate runs over the

Fig. 203.—GATE OPEN.

pulley and has a weight of iron or stone that nearly balances the gate. The opened gate is shown in figure 203.

Figure 204 shows a gate balanced in a similar manner,
and arranged so it can be opened by a person desiring to
drive through, without leaving the vehicle. It is suspended by ropes which pass over pulleys near the top of
long posts, and counterpoised by weights upon the other
ends of the ropes. Small wheels are placed in the ends
of the gate to move along the inside of the posts, and
thus reduce the friction. The gate is raised by means of
ropes attached to the center of the upper side of the
gate, from which they pass up to pulleys in the center of

the archway, and then out along horizontal arms at right angles to the bars which connect the tops of the posts. By pulling on the rope, the gate, which is but a trifle heavier than the balancing weights, is raised, and after the vehicle has passed, the gate falls of itself. In passing

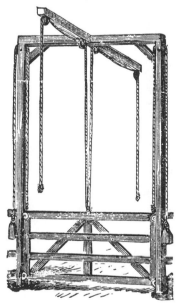

Fig. 204.—A " SELF-OPENING " GATE.

in the opposite direction, another rope is pulled, when the gate is raised as before.

Figures 206 and 207 show a gate specially designed for snowy regions. The latch-post, figure 205, is fixed in the ground and connected with the fence.   It is an ordinary square fence-post, to the side of which a strip of board is nailed, with a space of an inch between the board and the post.   At the opposite extremity of the gate a heel-post is set slanting, as shown in figures 206 and 207. The gate is made by laying the five horizontal bars on a barn floor or other level place, with one of the sloping cross-bars under them and the other above them.   Half inch

holes are bored through the three thicknesses, carriage
bolts inserted from below, and the nuts screwed on. The
gate, thus secured at one end, is carried to the place

Fig. 205.—LATCH-POST.   Fig. 206.—THE GATE OPEN.

where it is to remain and the other ends of the horizontal bars secured to the heel-post by similar bolts. These
should work freely in the holes. The lower bar is four

Fig. 207.—THE GATE CLOSED.

feet long and the upper bar seven feet. To the heel of
the upper bar is hung a weight nearly heavy enough to
balance the gate, so that it may easily be swung up, as
shown in figure 206, and the weight will keep it raised.

Figures 208 and 209 illustrate a very cheap way of making a hole through a picket fence in a place where there is not sufficiently frequent occasion for passing, to call for a more elaborate gate.   Strips of inch board, as

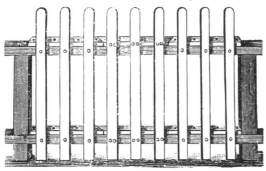

Fig. 208.—THE GATE IN POSITION.

wide as the rails of the fence, and five or six feet long, are nailed to the upper side of the rails and three pickets are nailed to the strips.   The pieces are then sawed off, beveling, and the pickets detached from the fence-bars

Fig. 209.—THE GATE OPEN.

by drawing or cutting the nails.   The gate can be lifted up and set at one side, but can not be pushed in or pulled out.   No rope or other fastening is required, besides it is almost invisible, which is many times an advantage. The gate, as lifted out of the fence and set on one side, is shown in figure 209.

Figure 210 shows an improved form of this gate without posts. In this case the small board strips are cut only as long as the gate is to be made wide, and a diagonal cross-brace running between them, as shown in the

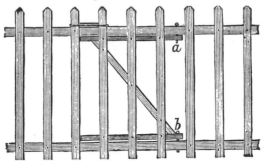

Fig. 210.—A SMALL GATE IN A PICKET FENCE.

engraving. The hinges are fastened to the horizontal bars of the fence by wooden pins shown at *a* and *b*. A piece of rope or a short wire passing over the ends of two of the pickets serves to keep the gate securely fastened. These openings are not designed for a regular gate, and

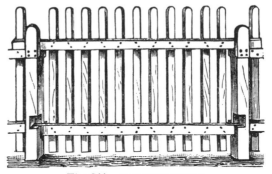

Fig. 211.—MOVABLE PANEL.

could not be used for the passage of any vehicle, as the horizontal bars would be in the way. For a back gate to the garden such an opening would frequently be found convenient and save many steps.

Figure 211 shows a lifting-gate, or rather, a movable

panel, wide enough to permit the passage of a team and vehicle. This might be useful in places where it was not desired to pass frequently.

Figure 212 shows another very convenient form of gate for use in a country where the snow is deep. It is fitted

Fig. 212.—A GATE NOT CLOGGED WITH SNOW.

in a strong frame, and is balanced by weights, so that it can be easily raised. The engraving sufficiently explains how this very useful gate is made and hung in the frame.

RUSTIC GATES.

A picturesque rustic gate is shown in figure 213. The fence and posts are made to correspond. Its manner of construction is clearly shown in the illustration. The vases on the top of the posts may be omitted, unless time can be taken to keep them properly watered.

A very neat, cheap, and strong rustic gate is shown in figure 214. The large post and the two uprights of the gate are of red cedar. The horizontal bars may be of the same or other wood. The longer upright is five and a half feet long, the shorter one four and a half feet.

The ends of the former are cut down to serve as hinges, as shown in the engraving. Five holes are bored through

Fig. 213.—ORNAMENTAL GATE.

each of the upright pieces, two inches in diameter, into which the ends of the horizontal bars are inserted and

Fig. 214.—LIGHT RUSTIC GATE.

wedged securely. For the upper hinge a piece of plank is bored to receive the gate, and the other end reduced and driven into a hole in the post, or nailed securely to its top. A cedar block, into which a two-inch hole has been bored, is partially sunk in the ground to receive the lower end of the upright piece. A wooden latch is in better keeping with the gate than an iron one.

BALANCE GATES.

Figure 215 is a modernized form of a gate which has for generations been popular in New England and the Middle States. In the primitive method of construction, the top bar consisted of the smoothly trimmed trunk of

Fig. 215.—BALANCE GATE.

a straight young tree, with the butt end projecting like a "heel" beyond the post upon which it turned. Upon its extremity a heavy boulder, or box of smaller stones, served as a counterweight. In the gate represented herewith the top stick is of sawn timber, upon the heel of which the large stone is held by an iron dowel. The other end of the top bar rests, when the gate is closed, upon an iron pin, driven diagonally into the post, as shown in the illustration. A smaller iron pin is pushed into the post immediately above the end of the top bar, to secure the gate against being opened by unruly animals, which may attempt to get in.

Figure 216 shows a balance gate which is used in some parts of North Carolina. It is a picket gate framed into

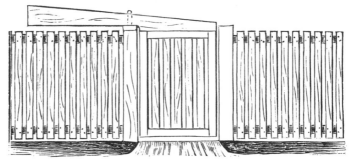

Fig. 216.—CAROLINA BALANCE GATE.

the lower side of a long pole, which is hung near its middle to a pivot driven into the top of the gate-post.

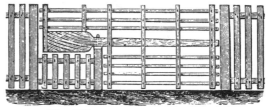

Fig. 217.—A TIDY BALANCE GATE.

Figure 217 shows a more elegant form, the "heel" of the gate remaining on a level with the top line of the fence.

GATE FOR SNOWY WEATHER.

The gate shown in figure 218 is suitable for all weather, but especially useful when there is a deep snow; for it is easily lifted up above the snow, and kept in place by putting a pin through holes in the hinge-bar, which is firmly fastened to the gate post. The hinge-bar should be of good, tough wood, and made round and smooth, so that the gate can swing and slide easily. Boards can be used in place of pickets if preferable. The latch-post to the

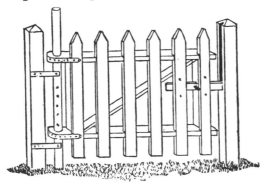

Fig. 218.—GATE FOR SNOWY WEATHER.

right, has a long slot for the latch to work in, instead of a hasp, so that it can be fastened when the gate is at any height.

WEST INDIA FARM GATES.

The illustrations, figures 219 and 220, show two forms of gates used on the island of Jamaica. These gates are

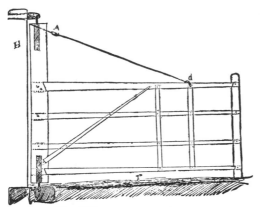

Fig. 219.—WIDE FARM GATE.

twenty-one feet long, each, and cannot possibly sag, even if any number of small boys swing on them. In gate figure 220 the main post is nine by six inches; the bars—marked 2, 3, 5 and 7—are let in the wood three inches on the upper side and one and a half inches on the lower. The tenons, indicated by the dotted lines, go entirely through the posts, and are fastened with pins. Brace 6 is attached to the upper bar eighteen inches beyond the center, F; D is a stout fence wire

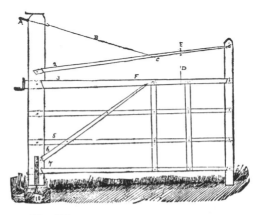

Fig. 220.—ANOTHER WIDE FARM GATE.

fastened by a screw nut at E; the wire, B, is held tightly by the screw hook, A; the iron band, 9, is an inch thick and is bolted to the post. It works on a pivot one and a quarter inches in diameter, and which turns on a flat piece of iron at the bottom of a piece of a one and a half inch iron pipe, which is soldered with molten lead in the stone, 10. Only hard wood is used in the construction. In the gate shown in figure 219, the construction differs from the one just described in that it has a light chain fastened in the shackle, C, and is screwed firmly at A. It is attached to the post, H, by a pivot, as seen in our illustration.

GATE HINGES OF WOOD.

It is often convenient and economical, especially in newly settled regions, where blacksmiths and hardware stores are not at hand, to supply hinges for gates, to make them of wood. The simplest and most primitive form is shown in figure 221. A post is selected having a large limb standing out nearly at right angles. A perpendicular hole in this secures the top of the rear gate standard. The foot rests in a stout short post, set against the main

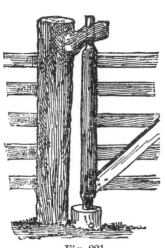
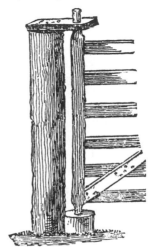

Fig. 221.   Fig. 222.

post. A small gimlet hole should extend outward and downward from the lowest side or point in the hole in the short post, to act as a drain, or the water collecting in it would be likely to soon rot both the standard and the short post itself. Another form is to hold the top by a strong wooden withe. A third form is illustrated in figure 222, in which the top of the standard passes through a short piece of sawed or split plank, spiked or pinned upon the top of the post.

The form shown at figure 223 is made of a stout lithe sapling or limb of beech, hickory or other tough hard wood or, if it is attainable, a piece of iron rod.

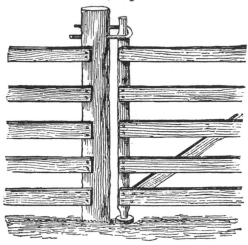

Fig. 223.—A WITHE HINGE.

A gate can be made without hinges by having the hanging stile somewhat longer than the front stile, and making both ends rounded. The lower one is to work

Fig. 224.—GATE WITHOUT HINGES.

in a hole in the end of a short post raised so that the soil will not readily get in, and the upper one works in a hole made in an oak piece attached to the top of the gate post. Gates of this kind can be made and hung with but little more expense than bars, and will be found far more convenient and saving of time than the latter.

Figure 225 represents a small hand-gate hung upon an iron pin driven into a hole bored in the bottom of the hinge-post, and one of similar size and material bent to a

Fig. 225.—SOCKET HINGES.

sharp angle, and fitted in the top. The lower pin rests in the sill and the upper one extends through the post to which the gate is hung.

DOUBLE GATES.

Figure 226 shows a substantial method of hanging two gates to the same post. The post may be of masonry

Fig. 226.—A DOUBLE GATE.

and the hinge bolts pass through the post, thus preventing any sagging. It is frequently convenient to have gates in the barnyard hung in this manner, then yards

Fig. 227.—DOUBLE BALANCE GATE.

may be shut off one way or the other by simply swinging the gates.

Figure 227 represents a balanced gate for a double driveway. The total length is thirty feet—sixteen feet on one

Fig. 228.—DOUBLE BALANCE GATE WITH STONE POST.

side of the supporting post and fourteen feet on the other. The horizontal top-piece may be of sawn timber, or better still, of a round pole cut from a straight young tree, the larger end being on the short side, its additional thickness serving to counterbalance the longer extremity

of the gate. The vertical strips of the original gate, from
which the sketch was made, were lag-sticks from an old
tread horse-power, and the chain was a part of the remains of a worn-out chain pump.   It is held in place by
staples driven into the vertical
pieces, as shown in the illustration.   A pin pushed into the post
at either end of the large top bar
fastens it securely when closed.

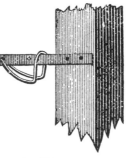

Figure 228 is a gate which combines some of the features of the
preceding two.   The stone pillar
is round, three feet across and four
and a half feet high.   A post is
placed in the center, upon the end

Fig. 229.—THE GATE
LATCH.

of which the bar rests, bearing the two gates.   The fence
is arranged in a sweeping curve, so that only one passage-
way can be open at once.

Figure 230 shows a style of double gate, which has

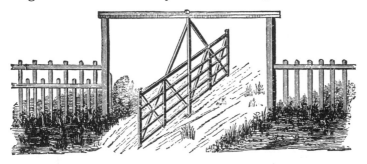

Fig. 230.—A DOUBLE HINGELESS GATE.

been found very useful on large stock farms, where it is
necessary to drive herds of cattle through it.   Two high
posts are set in the ground about twenty feet apart, and
a scantling is put on, which extends from the top of one
post to that of the other.   A two-inch hole is bored in
the center of this scantling, and a similar hole in a block

of wood, planted firmly in the ground in the center of the gateway. The middle post of the gate frame is rounded at each end to fit these holes, and this post is the pivot on which the gate turns. With this gate one cow cannot block the passage, besides there is no sagging of gate posts, as the weight of the gate is wholly upon the block in the center. To make the latch, figure 229, a bar of iron one and a half inch wide and eighteen inches long is bolted to one of the end uprights of the gate, and a similar bar to one of the posts of the gateway. For a catch, a rod of three-eighth inch iron passes through a half-inch hole near the end of the bar upon the gateway. This rod is bent in the form shown in the engraving, and welded. It will be seen that the lifting of this bent rod will allow the two bars to come together, and when dropped it will hold them firmly.

DOUBLE-LATCHED GATES.

Figure 231 represents a substantial farm gate with two latches. This is a very useful precaution against the

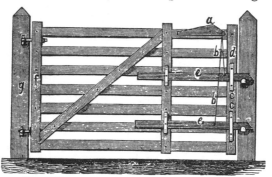

Fig. 231.—A DOUBLE-LATCHED FARM GATE.

wiles of such cattle as have learned to unfasten ordinary gate-latches. The latches work independently of each other, the wires, *b, b,* being fastened to the hand lever *a,* and then to the latches *e, e.* A roguish animal will sometimes open a gate by raising the latch with its nose, but if one attempt it with this, it can only raise one latch at a time, always the upper one, while the lower one remains fastened. As soon as the animal lets go, the latch springs back and catches again. A hog cannot get through, for the lower latch prevents the gate from opening sufficiently to allow it to pass. A cow will find it difficult to open the gate, because she cannot raise the gate high enough to unlatch it. The latches *e, e,* work up and down in the slides *c, c,* and when the gate is fastened they are about half-way between the top and bottom of the slides.

Figure 232 shows another form of double latches, which are closed by absolute motion, instead of depend-

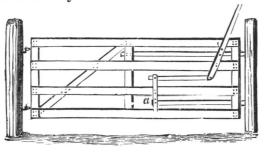

Fig. 232.—A GATE FOR ALL LIVE STOCK.

ing upon their own weight. There are two latches fastened to a jointed lever, so that when the upper end or handle is pushed backward or foward, the latches both move in the same direction. The construction of the gate, and the form and arrangement of the latches and lever, are plainly shown.

IMPROVED SLIDE GATE.

The old style slide gate is an unwieldly contrivance, and the only excuse for its use is its simplicity and cheapness. Numerous devices have been invented and

patented to make it slide easier and swing easier, but
their cost has prevented them from coming into general
use, and the old gate still requires the same amount of
tugging and heaving to open and close it.

Figure 233 shows the attachment. The blocks at top
and bottom are hard wood, one inch and a quarter thick.
The two boards should also be of hard wood. Between
the boards are one or two small iron or hard wood wheels,
turning upon half inch bolts, which pass through both
boards. The bars of the gate run on these wheels. The

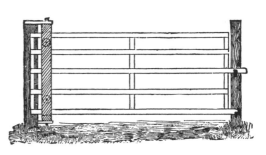

Fig. 233.　　　　　　Fig. 234.—THE GATE COMPLETE.

gate complete, with attachment, is shown in figure 234, the
gate being closed. To open the gate, run it back nearly
to the middle bar, then swing open. As the attachment
turns with the gate, the lower pivot should be greased
occasionally. It is well to fasten a barbed wire along the
upper edge of the top bar, to prevent stock from reaching over and bearing down on the gate. Where hogs are
enclosed, it is advisable to fasten a barbed wire along the
lower edge of the bottom bar, as it keeps small pigs from
passing under, and prevents large ones from lifting the
gate up, or trying to root under.

A COMBINED HINGE AND SLIDING GATE.

The illustrations, figures 235 and 236, show a gate very
handy for barnyards. It is fourteen feet wide for ordi-

nary use, and has three short posts. The middle one is movable. A box of two inch boards made to fit the post is planted in the ground ; in this the post is set, and can be removed at pleasure. This post is placed three feet from the outside one. The hinge is made of hard wood,

Fig. 235.—THE GATE OPEN.

with a wheel six inches in diameter, as shown in the engraving. It should be so constructed that the gate will move freely, but not too loosely. It is supported at the top by a cap, placed diagonally across, and at the bottom by a block of locust or cedar under it. The middle up-

Fig. 236.—THE GATE CLOSED.

rights of the gate should be placed a little to one side of the center, so that the gate can be balanced under the roller. Wooden catches are placed in the middle post, upon which the gate rests. To open the gate, push it back to the middle post, elevate the gate slightly, and it will roll down to the center, where it can be readily opened. Figure 235 shows the gate open, and in figure

236 it is seen closed. This gate has no latch. A barn-yard gate is not usually opened wide. A space large enough to admit a man or horse is all that is necessary in most cases. It is more easily opened than the ordinary gate, and it will stay where it is placed. By cutting a notch in the third board, and elevating it to the upper catch on the middle post, a passage is made for hogs and sheep, excluding larger animals.

GATES OF WOOD AND WIRE.

One of the cheapest and most popular styles of farm gate is made of plain or barbed wire, supported by

Fig. 237.—A NEAT GATE OF SCANTLING AND WIRE.

wooden frames. Figure 237 shows a very neat form of combination gate. To make it, obtain three uprights, three inches by one and a half inches, five and a half feet long, and four strips, three inches by one inch, eleven feet long. Cut shoulders in the ends of the strips, and saw out corresponding notches in the uprights;

make these one and a half inch, or half the width of the strips. The bottom notch is two and a half inches from the end of the upright, and the upper one nine and a half inches from the top end. Fit the strips into the notches. There is then a space of one inch between the strips, into which put inch strips, so as to make all solid, and fasten together with carriage bolts. Braces three by one and a half inches are inserted, and held in place by bolts or wrought nails. Bore as many holes in the end-pieces for one-quarter inch eye-bolts, as it is desired to have wires. Twist the wire firmly into the bolts on one upright, and secure the other end to the corresponding bolts on the upright at the opposite end. In stretching the wires, pass them alternately on opposite sides of the center piece, and fasten in place by staples. This will, in a measure, prevent warping. By screwing down the bolts with a wrench, the wires may be drawn as tightly as desired. The hinges are to be put on with bolts, and any sort of fastening may be used that is most convenient. Barbed or smooth wire may be used.

A GOOD AND CHEAP FARM GATE.

Figure 238 shows a gate of common fence boards and wire, which can be made by any farmer. The longer upright piece, seven feet long, may be made of a round stick, flattened a little on one side. The horizontal bars are of common fence boards cut to the desired length, and the shorter, vertical piece may be made of scantling, two by four inches. Three wires, either plain or barbed, are stretched at equal intervals between the upper and lower bar. A double length of wire is extended from the top of the long upright to the opposite lower corner of the

gate.   A stout stick is inserted between the two strands
of this diagonal brace, by which it is twisted until it is

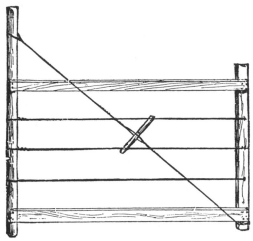

Fig. 238.—GOOD AND CHEAP FARM GATE.

sufficiently taut.   If the gate should at any time begin
to sag, a few turns brings it back.

AN IMPROVED WIRE GATE.

Figure 239 shows an improved form of wire farm
gate, in which the wires can be made tight at pleas-

Fig. 239.—IMPROVED WIRE GATE.

ure.   Instead of attaching the wires to both of the end
standards of the gate, a sliding standard is put on

near the end, to which the wires are fastened. This is
secured to the main standard by two long screw bolts,
leaving a space between the two of five or six inches.
The wires are tightened by turning up the nuts.

A plainer but very effective gate is shown in figure 240
The uprights are three and one quarter by two inches,
the horizontals twelve or thirteen feet long, by three and
a half by two inches, all of pine. The horizontals are
mortised into the uprights, the bolts of the hinges
strengthening the joints. The barbed wires prevent ani-

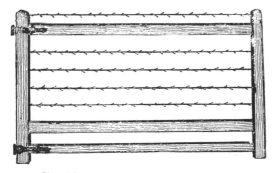

Fig. 240.—GATE OF WOOD AND WIRE.

mals from reaching over and through the gate. To put
in and tighten the wires, bore a three-eighth inch hole
in the upright, pass the wires through, one or two inches
projecting, plug up tightly with a wooden pin, and bend
down the ends of the wire. Measure the distance to the
other upright, and cut the wire two inches longer. Pass
the wire through the whole and tighten with pincers.
When the wire is stretched, plug up with a wooden pin,
and then bend down the wire. If the wire stretches, it
can be tightened very easily.

Figure 241 represents a light gate, that a child can
handle, which does not sag or get out of repair, and is
cattle proof. The materials are two boards, twelve or
fourteen feet long, three uprights, the end piece three

and one-half feet and the center four and one-half feet.
two strands of barbed wire, one between the boards, and

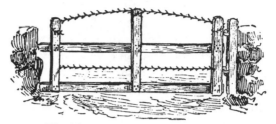

Fig. 241.—BARBED WIRE IN A GATE.

the other at the top of the uprights. It is hung the
same as the common form of gate.

TAKING UP THE SAG IN GATES.

Various means have been devised for overcoming the
sagging of gates. In figure 242 the hinge-post of the

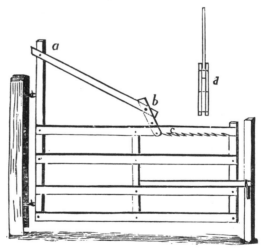

Fig. 242.—REMEDY FOR A SAGGING GATE.

gate-frame extends somewhat above the upper bar of the
gate. A board is fastened to the top of this post, *a*,

which runs downward to *b*, near the middle of the upper cross-bar, and then connects with a short double band— one on each side of the long board—which is provided with a bolt fitting into notches, *c*, cut in the under side of the upper bar of the gate. The form of the double-latch piece, with its bolts, and its attachment to the board is shown at *d*.

Figure 243 represents an arrangement which not only provides for taking up the sag, but also for raising the gate above encumbering snow. The gate is made of or-dinary inch boards put together with carriage bolts, upon which the joints play freely. The end of the gate, *a*, is made of two boards, and the post, *b*, is four by six inches.

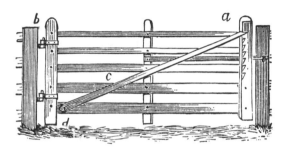

Fig. 243.—A LIFT-BAR FOR A GATE.

One board of the end, *a*, is notched. The diagonal piece, *c*, is fastened at *d*, by means of a bolt through it and the lower board. The end, *a*, of the diagonal piece, is shaped to fit the notches, by means of which the gate can be raised and lowered. It can also be used as a passage for pigs between fields, by simply raising the gate suffi-ciently to let them go through. A board, not shown in the engraving, is tacked to the notched board, to prevent the diagonal piece from slipping out of its place.

A much firmer gate is shown in figure 244. The hinge-post is about twice the height of the gate, and has a cap-piece, *a*, near the top. This cap is of 2 by 6 hard-

wood, strengthened by two bolts, *e, e,* and held in place
by two wooden pins, driven just above it and through

Fig. 244.—A REMEDY FOR A SAGGING GATE.

the tenon end of the post.   Wedges *c* and *d* are driven
in the cap on each side of the post.   Should the gate
sag, the wedge, *d,* may be loosened, and *c* driven further
down.   The lower end of the gate turns in a hole bored
in a hard-wood block placed in the ground near the foot
of the post.

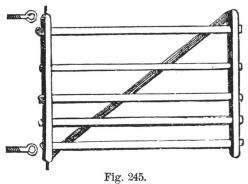

Fig. 245.

Figure 245 shows a gate similarly hung on pivots
driven into the ends of the hinge-bar.   These play in eye-

bolts which extend through the post to which the gate is hung, and are fastened by nuts on the other side. As the gate sags, the nut on the upper bolt is turned up, drawing the upper end of the hinge-bar toward the post, and lifting the gate back to a horizontal position.

GOOD GATE LATCHES.

Some cows become so expert, they can lift almost any gate latch. To circumvent this troublesome habit, latches made as shown in figure 246 will fill this bill exactly. It is a piece of iron bar, drawn down at one

Fig. 246.—GATE LATCH.

Fig. 247.—SPRING GATE CATCH.

end, and cut with a thread to screw into the gate post. A stirrup, or crooked staple, made as shown, is fitted by a screw bolt and nut to the bar. A small bolt must be driven in to keep the stirrup from being thrown over. A projecting slat on the gate, when it is shut, lifts the stirrup and holds the gate. This latch is too much for breechy cows, and they are never able to get "the hang of it."

A simple catch for a gate may easily be made from a

piece of seasoned hickory, or other elastic wood, cut in
the shape as shown at *a* in figure 247. This is fastened strongly to the side of the gate, with the pin, *c*,
working through the top loosely, so that it will play
easily. The catch, *b*, is fastened to the wall or post, as
the case may be. The operation will be easily under-
stood from the illustration, and it will be found a serviceable, sure, and durable contrivance. The gate cannot be
swung to without catching, and it may swing both ways.

A very simple and convenient style of fastening is illustrated in figures 248 to 251. It can be made of old

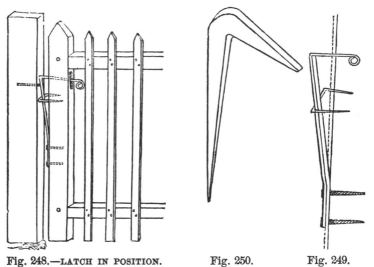

Fig. 248.—LATCH IN POSITION.   Fig. 250.   Fig. 249.

buggy springs, or any flat steel, and should be one inch
broad by three six-tenth inch thick, and about eighteen
inches long, at the distance of four inches from the lower
end. The lever is slightly bent, and has two screw or
bolt holes for fastening, figure 249. Eight inches of the
top portion is rounded and bent at right angles. The
upper part passes through a narrow mortise in the head-
post of the gate figure 248. A flat staple, large enough
to go over the spring holds it in place. An iron hook,

figure 250, driven into the post, holds the latch. A wooden lever, bolted to the top board of the gate, figure

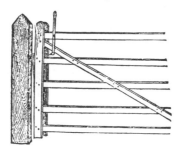

Fig. 251.—LATCH WITH TOP LEVER.

251, enables a person on horseback to open or close the gate. This latch can be applied to any kind of a gate, and is especially desirable in yards or gardens, when, by the addition of a chain and weight, one may always feel that the gate is securely closed. The latch does not cost more than fifty cents, and if properly made and put on will last as long as the gate.

Fig. 252.—GATE LATCH.

In figure 252 is represented a style of gate latch in use in some Southern States. It possesses marked advan-

tages, for certain purposes, over others.   It holds to an
absolute certainty, under all circumstances, and by allow-

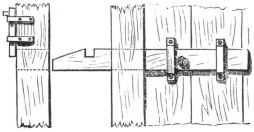

Fig. 253.—LATCH AND PIN.

ing the latch pin to rest on the bottom of the slot in the
post, it relieves the hinges and post from all strain. The
latch may be formed by a common strap-hinge, made to
work very easily, and the pin should be either a strong
oak one or an iron bolt or "lag screw."

Figure 253 shows a latch which cannot be opened by

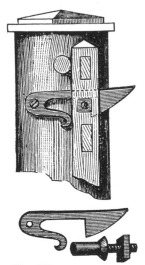

Fig. 254.—GATE LATCH.

the most ingenious cow or other animal.   The latch of
wood slides in two iron or wood bands screwed to the

gate. It is moved by a knob between the bands, which also prevents it from going too far. The outer end is sloping and furnished with a notch. It slides through a mortise in the gate post, indicated by dotted lines. When the gate is closed, the latch is slid through the mortise, and the drop-pin, which plays vertically in two iron bands, is lifted by the slope on the latch, and drops into the notch. It can be opened only by lifting the drop-pin, and sliding back the latch at the same time.

Figure 254 shows a very ingenious and reliable form of latch. The curved tail must be thin enough and sufficiently soft to admit of bending, either by a pair of large pincers or a hammer, just so as to adapt it to the passage of the pin bolted through the front stile of the gate. As the gate closes, the latch lifts out and the tailpiece advances. The catch-pin cannot possibly move out, unless the whole end of the gate moves up and forward.

TOP HINGE OF FARM GATE.

Continual use, more or less slamming, and the action of the weather, make the gate settle somewhat, but

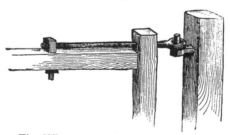

Fig. 255.—TOP HINGE OF FARM GATE.

the illustration, figure 255, shows a hinge which obviates this trouble. The upper hinge is made of a half-inch rod, about sixteen inches long, with an eye on one end, and a long screw-thread cut upon the other. This thread works in a nut, which nut has a bolt shank and nut, whereby it is firmly attached to the top bar of the gate. If the gate sags at all, it must be simply lifted off the thumbs, and the hinge given a turn or two in the nut; and the same is to be done in case of subsequent sagging. The hinge bolt must, of course, have some opportunity to move in the stile, and must be set long enough at first to allow the slack to be taken up whenever found necessary.

GATEWAYS IN WIRE FENCE.

Regular posts and bars at a passage-way through a wire fence are inconvenient and unsightly. A good sub-

Fig. 256.—GATEWAY IN A WIRE FENCE.

stitute for a gate is illustrated in figure 256. Light galvanized iron chains have a "swivel" near the end, by which they may be loosened or tightened, so as to be of

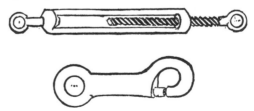

Figs. 257 and 258.—BUCKLE AND SNAP HOOK FOR CHAIN GATE.

just the right length, and a snap-hook at the other. These are both shown of larger size in figures 257 and 258.

The chains are attached by screw-eyes to the posts, and should correspond in number, as well as in position, with the wires. Thus they appear to be a continuation of the same, and as they are larger, they appear to the animals to be stronger, and even more dangerous than barbed

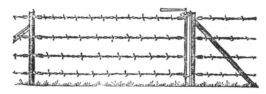

Fig. 259.—THE GATE CLOSED.

wire—hence are avoided. A short rod of iron may be made to connect them at the hook-ends, and so in opening and closing the way, they may all be moved at once.

A cheaper and simpler form of wire gate is shown in figures 259 and 260. It consists of the same number of strands as in the adjoining fence, attached to a post in the ordinary way at one end, while the other wire ends are secured to an iron rod. This rod is pointed at the lower end, and when the gate is closed, as seen in figure

Fig. 260.—THE GATE OPEN.

259, this end passes down through a loop, and the upper end is secured to a hook. In opening the gate, the rod is loosened and swings out, when the sharp end is thrust into the earth, or a hole in a wooden block set in the ground at the proper place to receive it.

Figure 261 shows a somewhat similar arrangement. The gate wires are fastened to one post with staples, and attach the loose ends to a five-foot pole. To shut the gate, take this pole or gate-head and put the lower end

Fig. 261.—A WIRE GATE.

back of the lower pin, and spring the upper end behind the one above. If the wires are all of the right length, they will be taut and firm. Two slats fastened to the gate wires will keep them from tangling. A short post set at one side of the gateway may be found convenient to hold the gate when open.

CHAPTER XII.

WICKETS AND STILES.

IRON WICKETS.

Wickets and stiles are convenient passageways through or over fences crossing foot-paths. The bow wicket has the advantage of providing a gate "always open and always shut," and not apt to get out of repair. A wrought iron bow wicket, with short vertical bars, is shown in figure 262. Figure 263 has the bars horizon-

tal, and folds in the middle for a wheel-barrow or small
animals to pass. To go through it, a person simply steps

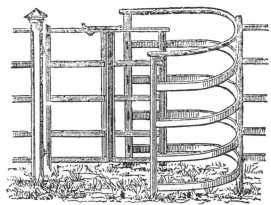

Fig. 262.—WICKET WITH HINGE.

into the bow, swings the gate away from him, and swings
it back in passing out.   There is no latch to fasten, and

Fig. 263.—WICKET WITH UPRIGHT BARS.

no fear of the entry of live stock.   Similar wickets may
be constructed of wood for board fences.

WOODEN WICKETS.

Figure 264 shows a wicket gate common in England,
where it is much used in foot-paths across fields, etc.   It

is an ordinary small gate, which swings between two posts, set far enough apart to permit the passage of a person.   These two posts are the two ends of a V-shaped

Fig. 264.—A GATE FOR FOOT-PATH.

end in the fence.   The engraving shows the construction of the end of the fence, with the two posts, between which the gate swings.

Figure 265 is another form of gate, which consists of a V-shaped panel, filling the opening in the fence—the open

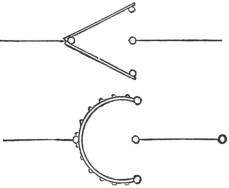

Figs. 265 and 266.—COMMON AND IMPROVED WICKETS.

ends of the V being fixed to posts equally distant from and in a line with one of the posts in the fence, and at right angles to it.   This is improved by using bent

wheel-rims, figure 266, instead of the straight pieces forming the V-shaped panel. Kept well painted, the hickory
rims will bear the exposure to the weather perfectly. The
palings should be of oak, an inch wide and half an inch
thick, fastened on with screws. The opening in these
stiles must be sufficient to allow a corpulent person to
pass easily, even if a frisky bull is in uncomfortable prox-

Fig. 267.—A CONVENIENT STILE.

imity, and for this figure 266 is really the most convenient form. The objection to both of these stiles is, that
there is no actual closing of the passage. Calves, sheep
and pigs, not to mention dogs, work their way through.
To prevent this, the gate-stile, figure 267, was invented.
It has a small gate swinging on the middle post, but
stopped in its movement by the end posts of the V. A

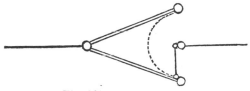

Fig. 268.—A GATE STILE.

person can pass by stepping well into the V and moving
the gate by him, where he has free exit. This form is
efficient, but inconvenient. A fourth form, the best of
all, is the swinging A-stile, figures 268 and 269. In this
there are two light gates, made upon the same hinge-
post, spreading like the letter A, and braced with a cross-
piece between the rails of each side, like the center part

of the A. This gate is set to swing on each side of the
center-post, as shown. It is so much narrower than the
V-stiles, that it is almost impossible for small animals to
pass, but it is easily hung so that it will always remain

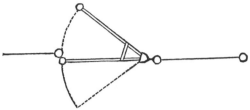

Fig. 269.—SWINGING STILE.

closed, and so offer no temptation to animals on the outside. At night, or when not in use, a wire ring or withe-
hoop thrown over the top of the post and the upright
part of the gate frame, will securely fasten it. To make

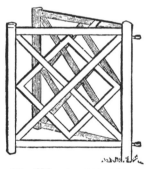

Fig. 270.—A NEAT GATE.

the gate swing shut, all that is necessary is to set the eye
of the lower hinge of the gate well out towards the outside. In figure 270 we give a neat A-gate, made of pine
or any strong and light wood.

———·◇·———

STILES FOR WIRE-FENCES.

The extensive use of wire-fences calls for a farm convenience, heretofore but little known in this country—the

stile. The manner of constructing one suitable for barb-
wire fence is shown so plainly in the engraving, figure 271,
that no description is necessary.   The cross piece, upon

Fig. 271.—STILE FOR BARB WIRE FENCE.

which one passes from one flight of steps to the other,
may be of any desired width.

Stiles of convenient forms for wire fences are shown in

Fig. 272.—FENCE STILE.

Fig. 273.—ANOTHER STILE.

figures 272 and 273.   The one seen in figure 272 takes
less space on each side of the fence, but it is not so simple as that shown in figure 273.

Figure 274 shows a passageway in a wire fence, which

requires no climbing, and while it presents an effectual
barrier to large animals, is readily passed by any but very

Fig. 274.—WIRE FENCE PASSAGE.

corpulent persons. It originated and was patented in
England, but we believe there is no restriction on its
construction and use in this country.

CHAPTER XIII.

FENCE LAW.

FENCING OUT OR FENCING IN.

The common law of England, which to a large extent
became the law of the original States, bound no one
to fence his land at all. Every person is bound under
that law to fence his own cattle in, but not bound to
fence other cattle out. Every owner of domestic animals
is liable for injury committed by them on the lands of
others, even though the land was wholly unfenced. But
this feature of the English common law was not suited

to the conditions which surrounded the early settlers in any part of this country. So long as any region is sparsely settled, the amount of unoccupied land is so much greater than the occupied, that it is cheaper to fence stock out, than to fence it in. Hence the English common law in regard to fencing has been superseded by statute in many of the States. In others it has always remained in force, or has been revived by later statutes. There is such great diversity on this point in the statutes of the several States, that, to quote from Henry A. Haigh's excellent " Manual of Farm Law, " " every one having occasion to look up any point of law, should ascertain the statutory provisions concerning it from some official source. Do not depend upon this book or any other book for them, because they are liable to change, and do change from year to year ; but go to your town clerk or justice of the peace, and examine the statutes themselves."

DIVISION FENCES.

The legal obligations of adjoining owners to build and maintain division fences, rests entirely upon the statutes of the respective States, save in cases where long usage has created prescriptive rights, or special agreement exists. Such fences are to be built on the boundary line, the expense to be borne equally by the parties, or each one shall make and maintain half the fence. If they cannot agree, or either refuses or neglects to do his share, the statutes provide methods by which the matter may be determined. In some of the States, two or more public officers, called fence-viewers, are elected annually in each township, whose duties, as prescribed by statute, are, when called upon, to hear and decide questions relating to fences in their respective towns. In other States, these duties are performed by overseers of highways or selectmen, *ex-officio*. Whenever any owner or occupant of land refuses to build or maintain half the division fence, or cannot agree with his adjoining neighbor as to which portion they shall respectively maintain, the fence-viewer may be called. Upon being so called, the fence-viewer shall upon reasonable notice, and after viewing the premises, determine and assign the respective portions of the fence to be maintained by each. The assignment when so made and recorded by the proper officer, becomes binding upon the present and all subsequent owners of the land. (2 Wis. 14). When by reason of a brook, watercourse, or natural impediment, it is impracticable or unreasonably expensive to build a fence on the true line between adjacent lands, and the owners thereof disagree respecting its position, the fence viewers may, upon application of either party, determine on which side of the true line, or whether partly on one side and partly on the other, and at what distances, the fence shall be built and maintained, and what portions by either party, and if either party refuses or neglects to build and maintain his part of the fence, the other shall have the same remedy as if the fence were on the true line. When a division fence shall be suddenly destroyed or prostrated by fire, winds or floods, the person who ought to repair or rebuild the same should do so in ten days after being notified for that purpose, and in the meantime he will be liable for damages done by estrays.

There is no legal obligation in any of the States, upon any proprietor of uncultivated, unimproved and unoccupied land, to keep up division fences. When a proprietor improves his land, or encloses land already improved, the land adjoining being unimproved, he must make the whole division fence, and if the adjoining proprietor afterward improves his land, he is required to pay for one half the division fence, according to the value thereof at that time. The laws of the respective States are not uniform touching the obligations to maintain one half a division fence after the owner of the land ceases to improve it. In Rhode Island and some other States, the proprietors are required to maintain these respective proportions, whether they continue to improve their land or not. In Maine, New Hampshire, Vermont and several other States, it is provided that if one party lays his lands common, and determines not to improve them, he may, upon giving due notice, cease to support such fences. But in most of the States, he must not take away any part of the division fence belonging to him and adjoining the next enclosure, provided the other party will allow and pay for his part of such fence. If the parties cannot agree as to its value, it may be decided by two or more fence-viewers. Where adjacent land is owned in severalty and occupied in common, and either party desires to occupy his in severalty, and the parties disagree, either party may have the line divided by the fence-viewers, as in other cases.

Owners of adjoining lands may agree between themselves as to the building and maintenance of division fences, and such agreements are valid, whether they are in accordance with the law or not. In some States such an agreement, if in writing, and filed with the clerk of the township, becomes binding upon all subsequent holders of the land. If not in writing, however, such an agreement may be terminated by either of the parties at pleasure.

HIGHWAY FENCES.

Under the common law, the land owner is under no obligation to fence his land along a public highway. But in Missouri, Iowa, Illinois, Oregon, and some other Western and Southern States, the common law rule has been modified by statutes depriving the land-holder of his action for trespass, unless he maintains sufficient fences around his land. In these States, the owner of land must enclose it with sufficient fences if he would cultivate it. Even where there is no such statutory provisions, it is practically necessary to maintain highway fences, as a protection against cattle which are driven along the highway. The use of barb wire for fencing along the public roads has given rise to questions for which there were no precedents. A case was decided in the United States Circuit Court, at Watertown, New York, December 17, 1885. The action was brought by a horse breeder to recover damages from his neighbor for injuries sustained by the plaintiff's horse from a barbed wire fence, stretched along the roadside in front of the defendant's premises. A non-suit was granted on the ground that the animal received the injuries through the contributory negligence of its owner. Among the rulings of the court was one permitting the plaintiff to be questioned, to show the fact that he had on his own farm a similar fence, but of sharper form of barb. The court further held that it might be a question whether it would not be competent testimony to show the common employment of barb wire fence in that region, and held that for the purpose of this case, a barbed wire fence, if properly constructed upon the highway, must be deemed a legal fence.

It may be said in a general way, that though there is no legal obligation resting on the land holder to maintain fences along the public highway, he neglects to do so at his own risk and peril.

WHAT IS A LEGAL FENCE?

What shall be necessary to constitute a legal and sufficient fence is specifically defined by the statutes of the several States, but there is no uniform rule among all. In Maine, New Hampshire, Massachusetts and many other States, it is provided that all fences four feet high,

and in good repair, consisting of rails, timber, boards, or
stone wall, and all brooks, rivers, ponds, creeks, ditches,
hedges, and other things deemed by the fence viewers to
be equivalent thereto, shall be accounted legal and sufficient fences.   In Vermont, Connecticut, Michigan, and
some other States, a legal fence must be four and a half
feet high.   In Missouri post fences must be four and one
half feet high, hedges four feet high, turf fences four
feet high, with ditches on each side three feet deep in the
middle and three feet wide; worm fences must be five
and one-half feet high to the top of the rider, or if not
ridered, five feet to the top of the top rail, and must be
rocked with strong rails, poles or stakes; stone or brick
fences must be four and one-half feet high.   In New
York the electors of each town may, by vote, decide for
themselves how fences shall be made, and what shall be
deemed sufficient.   No part of the fence law is so defi-
nitely regulated by the statutes of the respective States as
the requirements of a legal fence.   In all cases where
practical questions arise involving this point, it is best to
consult the statutes, which will be found in the office of
the township clerk.

RAILROAD FENCES.

In nearly every State, railroad companies are required
by statute to construct and maintain legal and sufficient
fences on both sides of their roads, except at crossings of
public highways, in front of mills, depots, and other
places where the public convenience requires that they
shall be left open.   The legal obligations of railroad companies to fence their roads rest wholly upon such statutes.
In New Hampshire it is provided that if any railroad
company shall neglect to maintain such fences, the owner
of adjoining land may build them, and recover double the
cost thereof of the company.   It is generally held by the
courts in all the States that, in the absence of such fences

the railroad company is liable for all resulting damage to live stock, and no proof of contributory negligence on the part of the owner of live stock is allowed as a plea in defence, the statute requiring such fences being a police regulation. When the railroad company has built a sufficient fence on both sides of its road, it is not liable for injuries which may occur without negligence on its part. If the fence is overthrown by wind or storms, the company is entitled to reasonable time in which to repair it, and if cattle enter and are injured, without fault on the company's part, it is not liable. If cattle stray upon the track at a crossing of a public road, and are killed, the owners cannot recover damages, unless the railroad company is guilty of gross negligence or intentional wrong. A law in Alabama making railroad companies absolutely liable for all stock killed on the tracks, was held to be unconstitutional.

CHAPTER XIV.

COUNTRY BRIDGES AND CULVERTS.

STRENGTH OF BRIDGES.

Bridge building is a profession of itself, and some of the great bridges of the world are justly regarded as among the highest achievements of mechanical science and skill. But it is proposed to speak in this work only of the cheap and simple structures for spanning small streams. The measure of the strength of a bridge is that of its weakest part. Hence, the strength of a plain wooden bridge resting upon timber stringers or chords, is equivalent to the sustaining power of the timbers in the middle of the span. The longer the span, other things being equal, the less its strength. The following table shows the sustaining power of sound spruce timber, of the dimensions given, at a point midway between the supports:

| LENGTH OF SPAN. | WIDTH AND THICKNESS OF TIMBER. | | | |
	6 by 8 inches.	6 by 9 inches.	6 by 10 inches.	6 by 12 inches.
Feet.	Pounds.	Pounds.	Pounds.	Pounds.
10	2,800	2,692	4,500	6,480
12	2,400	3,042	3,750	5,400
14	2,058	2,604	3,216	4,632
16	1,800	2,280	2,808	4,050

A stick of timber twenty feet between supports, will bear a load in its center only one half as great as a timber of the same dimensions, ten feet between supports. Thus four timbers six by twelve inches, in a span of sixteen feet, would bear a load of eight tons; in a twelve foot span, the same timbers would support a weight of nearly twelve tons.

BRACES AND TRUSSES.

The above is the initial strength of the timbers which support the weight of the superstructure, and any load that it may have to sustain. But in bridge building these timbers are reinforced by trusses or braces, which add greatly to the sustaining power of the bridge.

Figure 275 shows the simplest form of a self-supporting bridge, which will answer for spans of from ten to fifteen feet in length. The braces, c, c, reach from near

the end of the sill to about four feet above the center.
The truss rod, d, is one inch in diameter for short bridges

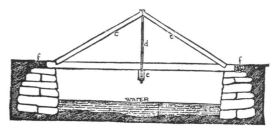

Fig. 275.—A SIMPLE FORM OF BRIDGE SPAN.

up to two inches for longer spans; it is provided with an
iron washer at the top.   The rod passes through the sill,

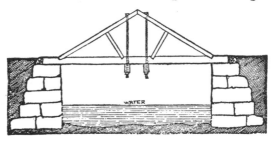

Fig. 276.—A STRONGER SPAN.

and a cross sill, e, which passes under the main sills, thus
adding firmness to the whole structure.   Logs, f, f, are

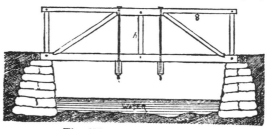

Fig. 277.—A SHORT BRIDGE.

placed against the ends of the sills to keep them in place,
and where the wheels will first strike them instead of the

floor plank, thus greatly equalizing the pressure. Figure 276 represents a modification of the above. The two truss rods and braces give the structure greater strength and solidity, adapting it for spans eighteen feet in length. For the latter length, sills should be of good

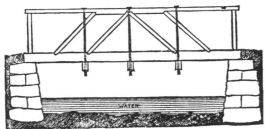

Fig. 278.—A BOLT TRUSS.

material, ten inches wide and fourteen inches deep, with three middle sills of about the same size.

Figure 277 is a more improved style of bridge, the truss serving both to support the structure, and as a parapet. The top railing is of the same width as the sill, about one foot. The lower side may be cut away, giving the bridge a more finished appearance. The railing at the center is six inches thick, and three inches at the ends. The tie, h, is full width and four inches thick. A bridge of this kind will answer for heavy traffic, even if twenty

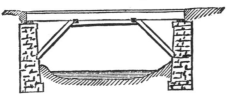

Fig. 279—BRIDGE BRACED FROM BELOW.

feet in length. The bolt truss, in figure 278, is adapted for a span of twenty-five feet. This makes a bridge of great firmness. Each set of truss-rods support a cross-sill. The road planks are laid crosswise of the bridge. The middle sills are sometimes half an inch lower than those along the sides, and should be four or five in number. The ends of the planks fit closely against the inside of the truss sills, thereby keeping the planks securely in place.

A common method of bracing is from below as shown in figure 279. This is not usually a good practice, as the braces are liable to be carried away by ice or floods.

ABUTMENTS, PIERS AND RAILINGS.

If the sills of a bridge are laid directly upon the dry walls of an abutment, or upon a heavy plank, the jar of passing teams soon displaces some of the stones, and brings undue strain upon certain portions of the wall.

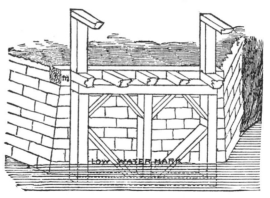

Fig. 280.—END OF A BRIDGE.

To avoid this, abutments are best made of cut stones, and laid in cement. A wooden bent for the support of the ends of the bridge may be made as shown in figure 280. The whole should be constructed of heavy timber, pinned together. A coat of white lead should cover the interior surface of all joints. The number and position of the posts of the wooden abutment are seen in the engraving. A log should be laid upon the wall at m, to re-

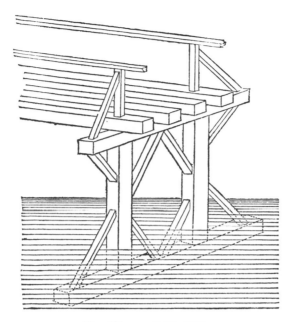

Fig. 281.—FRAMED PIER.

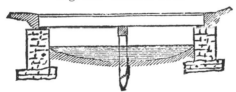

Fig. 282.—BRIDGE SUPPORTED BY PILES.

Fig. 283.   RAILING OF BRIDGE.   Fig. 284.

lieve the bridge from the shock of the passing wagons.
A center pier should be avoided as much as possible,
as it offers serious obstruction in floods, and ice, drift

wood and other floating matter become piled against it, seriously imperiling the entire structure. But in cases where the length of the bridge is so great as to require one or more piers, they may be constructed on the plan

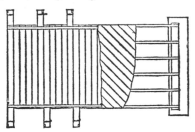

Fig. 285.—PLANK FLOOR OF BRIDGE.

shown in figure 281, or in case the bottom is so soft as to render the mudsill insecure, a line of piles supporting a cross-timber, as in figure 282. A strong, reliable parapet or railing should always be provided. The want of one may be the cause of fatal accidents to persons and horses. Figure 283 gives a side view of a good railing, and figure 284 shows the manner of bracing the posts to the ends of the cross-beams. They should be thus braced at every alternate post of the railing. The floor should be double, as shown in figure 285, the lower planks laid diagonally, and the upper layer crosswise.

─────◆◇◆─────

BRIDGES FOR GULLIES.

For small gullies which cross roadways or lanes in farms, and are not the beds of constant streams, but are occasionally filled with surface water, a very simple bridge is sufficient. One like that shown in figure 286 is as good as any. The sills, *a, a,* are sunk in a trench dug against the bank and at least to the level of the bed of the creek. The cross-sills, *b, b,* are not mortised into them, but simply laid between them. The pressure is

all from the outside, hence it will force *a, a,* tighter
against the ends *b, b,* which must be sunk a little into the
bed of the creek at its lowest point.   The posts are mortised into the sills, *a, a,* and plates, *c, c,* and *d, d,* upon

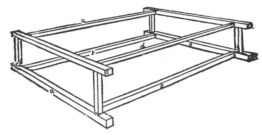

Fig. 286.—FRAME FOR BRIDGE.

which the planks are laid.   Props may be put against the
lower sides of the posts to hold the bridge against the
stream.

A cheap but practicable bridge is shown in figure 287.
Two logs are laid across the gully, their ends resting on
the banks, and to them puncheons or planks are spiked
to form the bridge. Stout posts, well propped and reaching above the highest water mark, are placed against the
lower side of the logs.   If the creek rises, the bridge,

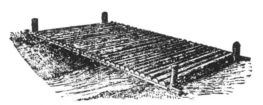

Fig. 287.—CONVENIENT FARM BRIDGE.

being free, will be raised on the surface of the water,
while the posts will prevent its being carried away.
Should it not rise with the water, it opposes so little surface to the current that the posts will hold it fast.

ORNAMENTAL BRIDGES.

No feature adds more to the appearance of ornamental grounds than tasteful bridges. A stream or narrow channel connecting two parts of a small sheet of water, affords an opportunity for the introduction of a bridge.

Fig. 288.—RUSTIC BRIDGE.

In the absence of such features a bridge may be thrown across a dry ravine. Whatever style may be adopted, should harmonize with the general character of the surroundings. An elaborate bridge of wood or masonry would be as much out of place on grounds unadorned by other structures, as a rude rustic one would be near highly

Fig. 289.—A BRIDGE OF ROCKS.

finished summer-houses and other architectural features. On most grounds a neat rustic bridge, something like the one shown in figure 288, would be in good keeping with its environments. Such bridges may be made of red

cedar logs and branches, resting upon stone abutments.
Where boulders are abundant, a stone bridge, something
like figure 289, may be built at very little cost, and will
last for generations.   The pleasing effect of rustic or
other ornamental bridges is enhanced by training Virginia creeper or other climbing plants upon them.

ROAD CULVERTS.

A culvert under a road is, in effect, a short bridge.
The simplest form of plank culvert, resting upon stone
abutments, is shown in figure 290.   Such a structure is
cheaply built, and serves a good purpose while the wood-
work remains sound.   But the planks wear out and the
timbers decay, requiring frequent renewing.   Where
stone is abundant it is much cheaper in the end to build
wholly of stone, as in figure 291.   After the abutments

Fig. 290.—CULVERT WITH PLANK FLOOR.

are built, a course of flat stone, along each side, projects
inward from six to ten inches, as at *a, a*, which are covered
with a broad stone, *b*.   Where the stream to be crossed is
so narrow that a row of single stones is sufficient to cover
the opening, a culvert like that seen in figure 292 is
cheaply made.   Such structures will remain serviceable
for a generation, if the foundations are not undermined
by the action of the water.

Where flat stones enough cannot be easily procured, culverts may be built of concrete. The abutments are first made, as in other cases; then empty barrels or sugar

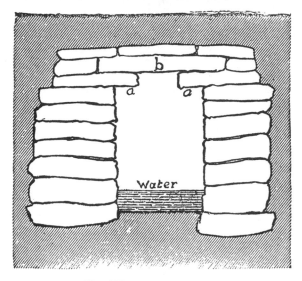

Fig. 291.—STONE CULVERT.

hogsheads, according to the capacity of the opening, are fitted in, or better still, a temporary arch is made of rough, narrow boards. The concrete of cement, sand

Fig. 292.—CHEAPER STONE CULVERT.

and gravel, is then prepared and poured in, temporary supports of lumber having been fixed across each end of the culvert to keep the concrete in place until it hardens.

Small stones may be mixed with the concrete as it is poured into place, and the whole topped off with a row of them. This protection of stones on the top is valuable, in case the covering of earth is worn or wasted away

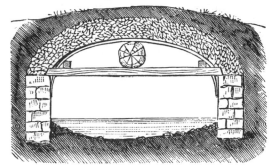

Fig. 293.—ARCHED CONCRETE CULVERT.

at any time while it is in use. For a longer culvert a flattened arch is made of concrete, as shown in figure 293. Light timbers are laid across, the ends resting lightly on the abutments. Across the middle of these a round log is placed to support the crown of the arch. Elastic split poles are sprung over all, and upon these are

Fig. 294.—ANGULAR CONCRETE CULVERT.

nailed thin narrow boards, extending lengthwise of the culvert. The ends being temporarily protected, the concrete is mixed and poured on, as before. As soon as the concrete has become thoroughly well "set," the light cross-sticks are cut in two and the temporary work removed.   A cross-section, showing another form of concrete culvert, and the method of construction, are shown in figure 294.   Such a culvert is more easily built than the last, but is not as strong.   The best and most durable culvert is of stone, with a regular half-round arch. Such work can only be done properly by a regular mason, but in the end it is cheaper, where the stone can be obtained, than any kind of make-shift.

INDEX